AF355995

Additive Manufacturing of Alloys and Composites

Additive Manufacturing of Alloys and Composites

Lehua Liu
Haokai Dong
Haizhou Lu
Chao Zhao

Basel • Beijing • Wuhan • Barcelona • Belgrade • Novi Sad • Cluj • Manchester

Lehua Liu
South China University
of Technology
Guangzhou
China

Haokai Dong
Chinese Academy of Sciences
Ningbo
China

Haizhou Lu
Guangdong Polytechnic
Normal University
Guangzhou
China

Chao Zhao
Huazhong University of
Science and Technology
Wuhan
China

Editorial Office
MDPI AG
Grosspeteranlage 5
4052 Basel, Switzerland

This is a reprint of articles from the Special Issue published online in the open access journal *Materials* (ISSN 1996-1944) (available at: www.mdpi.com/journal/materials/special_issues/HWLEHBJ461).

For citation purposes, cite each article independently as indicated on the article page online and using the guide below:

Lastname, A.A.; Lastname, B.B. Article Title. *Journal Name* **Year**, *Volume Number*, Page Range.

ISBN 978-3-7258-1832-7 (Hbk)
ISBN 978-3-7258-1831-0 (PDF)
https://doi.org/10.3390/books978-3-7258-1831-0

Contents

About the Editors

Lehua Liu

Lehua Liu, who holds a Ph.D. in Mechanical Engineering from Tsinghua University, is currently an associate professor at the School of Mechanical and Automotive Engineering at South China University of Technology. His primary research areas include the design and manufacturing of titanium alloys and amorphous alloys, die-casting materials and processes, high-throughput techniques, and machine learning.

Haokai Dong

Haokai Dong, who received his Ph.D. degree of Materials Science and Engineering from Tsinghua University, is currently an associate research fellow of the Ningbo Institute of Materials Technology and Engineering at the Chinese Academy of Sciences. His primary research areas include high-strength steels, copper alloys, and additive manufacturing.

Haizhou Lu

Haizhou Lu, who received his Ph.D. in Materials Science and Engineering from South China University of Technology, is currently an associate professor of the School of Mechatronic Engineering, Guangdong Polytechnic Normal University. His current research interests focus on fabricating high-performance Ti alloys and NiTi-based alloys by additive manufacturing.

Chao Zhao

Zhao Chao, who earned his doctorate in Materials Processing Engineering from South China University of Technology, is now a lecturer at the Department of Materials Science and Engineering at Huazhong University of Science and Technology, where he conducts research on various topics including advanced non-ferrous metal solidification, toughness mechanisms, and structure–function integration.

materials

Editorial

Additive Manufacturing of Alloys and Composites

Haokai Dong [1,2,*], Haizhou Lu [3] , Chao Zhao [4] and Lehua Liu [1,2,*]

1   National Engineering Research Center of Near-Net-Shape Forming for Metallic Materials, South China University of Technology, Guangzhou 510640, China
2   School of Mechanical and Automotive Engineering, South China University of Technology, Guangzhou 510641, China
3   School of Mechatronic Engineering, Guangdong Polytechnic Normal University, Guangzhou 510665, China
4   School of Materials Science and Engineering, Huazhong University of Science and Technology, Wuhan 430074, China
*   Correspondence: donghk@scut.edu.cn (H.D.); liulh@scut.edu.cn (L.L.)

Citation: Dong, H.; Lu, H.; Zhao, C.; Liu, L. Additive Manufacturing of Alloys and Composites. *Materials* **2023**, *16*, 992. https://doi.org/10.3390/ma16030992

Received: 10 January 2023
Accepted: 17 January 2023
Published: 21 January 2023

The emergence and development of high-performance materials have benefited from the revolution in modern manufacturing technology, in which additive manufacturing (AM) is the most representative over the last four decades. AM, also known as 3D printing, refers to a family of layer-upon-layer building technologies capable of producing geometrically complex engineering parts with a short lead time [1]. Compared to the conventional processing route, AM can provide more design freedom and flexible manufacturability, which has played an increasingly vital role in many custom fields such as patient-specific implants for medical application and the complex hollow structure of jet engine parts.

Nowadays, AM is widely used to fabricate metallic materials such as steels, nonferrous alloys, and high entropy alloys [2–6]. The repetitive high thermal gradient during AM leads to microstructural development which significantly deviates from the equilibrium condition, thereby resulting in copious metastable microstructures including hierarchically heterogeneous microstructures, multiphase constituent, nanosized precipitates and dislocation network, enabling extraordinary mechanical behaviors compared to the counterparts made by conventional methods. AM of metal matrix composites (originally invented to combine the unique properties of metals) is another hot research field which has grown in recent years [7]. AM, particularly powder-based AM methods, is proven to be a useful and versatile composite manufacturing technique that facilitates the processing of reactive primary powders for creating new materials with different constituent phases. Several intractable issues, such as weak interface bonding, cracks at the interfaces, inhomogeneous dispersions of reinforcement and residual stress induced by thermal mismatches between composing phases, appearing in conventional synthesis, could also be effectively mitigated and solved by AM.

In aiming to further develop additive manufactured alloys and composites, a full understanding of processing–microstructure–property relationships during AM is still the current scope for AM researchers, although great progress has already been made in respect of the large number of related papers published in recent years. The remaining critical challenges, including high production cost, the formation of various defects, and many inapplicable established theories in textbooks to explain physical metallurgy during AM, will continue to stimulate AM research [6], and we believe those challenges will be overcome in the near future.

In light of these contributions, the current Special Issue entitled "Additive Manufacturing of Alloys and Composites" welcomes these original research articles, state-of-the-art reviews, and perspectives on recent developments in additive-manufactured alloys and composites, which aims to provide analysis, solutions and support for creating more suitable materials for AM.

Conflicts of Interest: The authors declare no conflict of interest.

References

1. Gibson, I.; Rosen, D.W.; Stucker, B. *Additive Manufacturing Technologies*; Springer: Berlin/Heidelberg, Germany, 2014.
2. Gu, D.; Meiners, W.; Wissenbach, K.; Poprawe, R. Laser additive manufacturing of metallic components: Materials, processes and mechanisms. *Int. Mater. Rev.* **2012**, *57*, 133–164. [CrossRef]
3. Herzog, D.; Seyda, V.; Wycisk, E.; Emmelmann, C. Additive manufacturing of metals. *Acta Mater.* **2016**, *117*, 371–392. [CrossRef]
4. Manakari, V.; Parande, G.; Gupta, M.; Lopez, H.F. Selective laser melting of magnesium and magnesium alloy powders: A Review. *Metals* **2017**, *7*, 2. [CrossRef]
5. Bajaj, P.; Hariharan, A.; Kini, A.; Kürnsteiner, P.; Raabe, D.; Jagle, E.A. Steels in additive manufacturing: A review of their microstructure and properties. *Mater. Sci. Eng. A* **2020**, *772*, 138633. [CrossRef]
6. Haghdadi, N.; Laleh, M.; Moyle, M.; Primig, S. Additive manufacturing of steels: A review of achievements and challenges. *J. Mater. Sci.* **2020**, *56*, 64–107. [CrossRef]
7. Dadbakhsh, S.; Mertens, R.; Hao, L.; Van Humbeeck, J.; Kruth, J.P. Selective laser melting to manufacture "in situ" metal matrix composites: A review. *Adv. Eng. Mater.* **2019**, *21*, 1801244. [CrossRef]

Article

Study on the Effect of Microstructure and Inclusions on Corrosion Resistance of Low-N 25Cr-Type Duplex Stainless Steel via Additive Manufacturing

Yang Gu [†] , Jiesheng Lv [†], Jianguo He *, Zhigang Song, Changjun Wang, Han Feng and Xiaohan Wu

Research Institute of Special Steels, Central Iron & Steel Research Institute Co., Ltd., Beijing 100081, China; thiagoyoungkoo@163.com (Y.G.); lvllvlv@foxmail.com (J.L.); songzhigang@nercast.com (Z.S.); wangchangjun@nercast.com (C.W.); fenghan@nercast.com (H.F.); wuxiaohan@nercast.com (X.W.)
* Correspondence: hejianguo@nercast.com
[†] These authors contributed equally to this work.

Abstract: Duplex stainless steels are widely used in many fields due to their excellent corrosion resistance and mechanical properties. However, it is a challenge to achieve duplex microstructure and excellent properties through additive manufacturing. In this work, a 0.09% N 25Cr-type duplex stainless steel was prepared by additive manufacturing (AM) and heat treatment, and its corrosion resistance was investigated. The results show that, compared with S32750 duplex stainless steel prepared by a conventional process, the combination value of film resistance and charge transfer resistance of AM duplex stainless steel was increased by 3.2–5.5 times and the pitting potential was increased by more than 100 mV. The disappearance of residual thermal stress and the reasonable distribution of Cr and N elements in the two phases are the reasons for the improvement of the corrosion resistance of AM duplex stainless steel after heat treatment. In addition, the extremely high purity of AM duplex stainless steel with no visible inclusions resulted in a higher corrosion resistance exhibited at lower pitting-resistance-equivalent number values.

Keywords: additive manufacturing; duplex stainless steels; nano-inclusion; microstructure; corrosion resistance

Citation: Gu, Y.; Lv, J.; He, J.; Song, Z.; Wang, C.; Feng, H.; Wu, X. Study on the Effect of Microstructure and Inclusions on Corrosion Resistance of Low-N 25Cr-Type Duplex Stainless Steel via Additive Manufacturing. *Materials* **2024**, *17*, 2068. https://doi.org/10.3390/ma17092068

Academic Editor: Andrea Di Schino

Received: 15 March 2024
Revised: 18 April 2024
Accepted: 23 April 2024
Published: 28 April 2024

1. Introduction

Duplex stainless steels (DSSs) are composed of both ferrite and austenite phases, combining the excellent mechanical properties of ferrite with the superior corrosion resistance of austenitic stainless steel [1–3]. They are widely used in industrial and maritime applications [4–6]. DSSs can achieve theoretical corrosion resistance performance and actual service performance higher than the 300 series unitary austenitic stainless steel at similar or lower raw material costs [7]. The corrosion resistance of DSSs mainly depends on composition and microstructure.

AM technology is a production method that reduces costs and increases efficiency [8]. However, the relationship between the composition, process, microstructure, and properties of AM duplex stainless steel has not been systematically studied [9,10]. Research on the corrosion resistance of AM DSS steels are focused on utilizing the high forming temperature and rapid cooling rate characteristics of additive manufacturing to form numerous nano-sized oxide inclusions in the matrix [11,12]. Nano-inclusions bring benefits to the improvement of mechanical properties of duplex stainless steels, but the effect on the corrosion properties is not clear. Zhang et al. [13] investigated the substantial enhancement of mechanical properties in S32205 duplex stainless steel achieved through the incorporation of specialized nano-inclusions, with a powder oxygen content reaching 0.11 wt.%. Nano-inclusions bring benefits to the improvement of the mechanical properties of duplex stainless steels, but the effect on corrosion properties is not clear; while the measured corrosion potential increased, the pitting potential decreased. Haghdadi et al. [14] found that the corrosion resistance of AM 2205 duplex stainless steel was lower than that of hot rolled samples, and heat treatment was a necessary step to restore its pitting resistance; meanwhile, heat treatment was effective in restoring the duplex microstructure. Majld et al. [15] concluded that the decrease in the corrosion resistance of AM duplex stainless steel was caused by Cr_2N precipitation. As reported by other investigations [16–18], the corrosion resistance of AM DSSs is commonly comparable or inferior to that of traditionally manufactured counterparts.

On this basis, this paper primarily focuses on a 25Cr-7Ni low-N (0.09 wt.%) duplex stainless steel through an established manufacturing process. Its corrosion resistance before and after heat treatment was investigated and compared, with the conventional S32750 duplex stainless steel used as a reference. Meanwhile, a variety of characterization methods for microstructure and substructure were used to analyze the reasons for the improvement of corrosion resistance from the perspective of inclusions, element distribution, and microstructure, which provided a research direction for breaking through the limitation of composition on the corrosion resistance of duplex stainless steel and achieving higher corrosion resistance under the premise of lower theoretical corrosion resistance.

2. Materials and Methods

A 0.09% N 25Cr-type duplex stainless steel composition was designed. The low-nitrogen composition is designed to avoid the precipitation of various nitrides. Argon gas atomization was used to prepare 25Cr-type DSS powder, with a particle size distribution of 15–50 μm. The main chemical composition of the powder was measured via inductively coupled plasma atomic emission spectrometry (ICP–AES), and the results are shown in Table 1, with reference to the conventional preparation S32750 duplex stainless steel with a higher pitting-resistance-equivalent number (PREN) value, which is 42.1, while that of the 25Cr-type DSS is 38.6. Figure 1a shows the morphology observed under scanning electron microscopy (SEM) (supplied by FEI Co., Ltd., Hillsboro, OR, USA), revealing smooth, rounded particles without satellite powders. X-ray diffraction (XRD) analysis of the powder indicates a phase composition of 99.47% ferrite and 0.53% austenite, with no harmful phases.

Table 1. Chemical composition of the tested DSSs.

	Cr	Ni	Mo	N	Mn	O	Si	C	PREN
AM powder	24.70	6.52	3.74	0.098	0.55	0.028	0.35	0.0050	38.6
S32750	25.39	6.72	3.71	0.28	0.40	0.0022	0.43	0.020	42.1

Note: PREN = $W_{Cr} + 3.3 \times W_{Mo} + 16 \times W_N$.

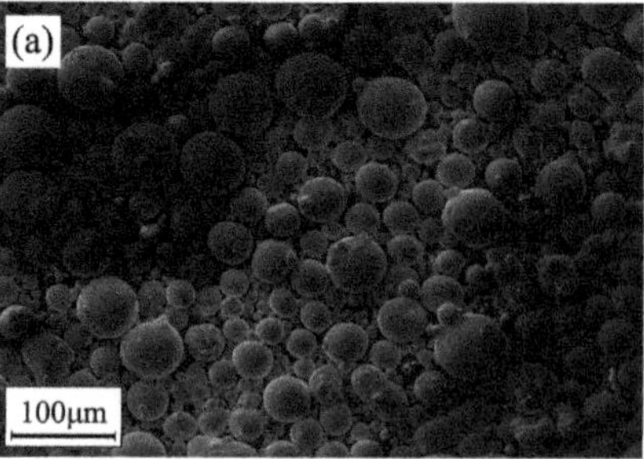

Figure 1. Powder SEM morphology: (**a**) 500× magnification; (**b**) 3000× magnification.

The aforementioned powder was processed using a DLM-280 metal selective laser melting (SLM) machine (supplied by Pera Corporation Ltd., Shanghai, China). The building process took place on a 316 stainless steel substrate in a high-purity argon atmosphere

(99.9%). The specific sintering parameters were as follows: laser input power (P) of 190 W, laser spot diameter of 0.1 mm, powder layer thickness (h) of 0.02 mm, line spacing (t) of 0.1 mm, scanning speed (v) of 850 mm/s, and a bidirectional scanning pattern with a 90° angle between each layer. The calculated energy density, obtained using Formula (1), was 117.65 J/mm^3. The density of the as-built product, measured using the Archimedes drainage method, was 7.81 g/cm^3, while the density of the conventional forged product was 7.82 g/cm^3, resulting in a relative density of 99.87% [19].

$$E = \frac{P}{v \cdot h \cdot t} \tag{1}$$

The heat treatment process was as follows: The AM samples were held at 1200 °C and 1100 °C for 1 h and then water-quenched. As a reference, the samples for the conventional process were held at 1100 °C for 1 h and then water-quenched.

The optical microscope (OM) specimens of $10 \times 10 \times 2$ mm were mechanically ground and polished using a diamond polish with a granularity of 5 μm and then immersed in a potassium permanganate–sulfuric acid aqueous solution at 50 °C for 3 h. The OM microstructure was observed using a LEICA MEF4M optical microscope (supplied by Leica Microsystems Shanghai Ltd., Shanghai, China). Polished samples were subjected to statistical analysis for the macroscopic distribution and quantity of inclusions using the ASPEX metal inclusion analyzer (supplied by FEI Co. Ltd., Hillsboro, OR, USA), with a scanning area of 7×7 mm.

Transmission electron microscope (TEM) specimens were mechanically thinned to a thickness of 40 μm and then mechanically punched to obtain circular disks with a diameter of 3 mm. Further thinning of the disks was performed using a dual-jet electropolisher at 28 V and −20 °C. The electrolyte consisted of 10% perchloric acid and 90% anhydrous ethanol. Observation was carried out using a FEI TECNAI G2 F20 (supplied by FEI Co. Ltd., Hillsboro, OR, USA) operated at an acceleration voltage of 200 kV. Electron probe X-ray microanalyzer (EPMA) experiments were conducted using a JXA-8530F PLUS (Electronics companies of Japan, Tokyo, Japan) electronic probe. The backscattered electron diffraction (EBSD) experiments were conducted using a FEI Quanta650 field emission scanning electron microscope (supplied by FEI Co. Ltd., Hillsboro, OR, USA). EBSD samples were immersed in a 10% alcoholic hydrochloric acid solution and subjected to electrolytic polishing at a voltage of 25 V for 30 s. EBSD characterization was carried out using a FEI Quanta650 field emission scanning electron microscope, and the data were processed using Channel 5 software. The EBSD data collection was conducted with a step size of 0.6 μm and a resolution of 400×400.

The electrochemical experiments were carried out using a standard three-electrode system, with the polished sample to be tested as the working electrode (WE), the platinum electrode as the counter electrode (CE), and the saturated calomel electrode (SCE) as the reference electrode. The electrochemical testing employs a 3.5 wt.% sodium chloride solution. Prior to the test, cathodic polarization was performed at −1 V vs. Ref. for 3 min to remove the passivation layer on the sample surface. Subsequently, an open-circuit (OC) voltage test was conducted for half an hour, allowing the voltage to stabilize within the range of ±10 mV before proceeding to the next step. The frequency range for electrochemical impedance spectroscopy (EIS) was from 10^{-2} Hz to 10^5 Hz. The impedance measurement signals had an amplitude of ±10 mV vs. OC sinusoidal waveforms, and the curve fitting was performed using ZSimpWin software (Version 3.6 EChem Software, Ann Arbor, MI, USA, http://www.echemsw.com (accessed on 2 March 2024)). For cyclic polarization curve testing, the initial voltage was set at −0.5 V vs. OC. The anodic current was scanned until it reached 100 μA, and then, a reverse scan was performed, terminating at the open-circuit potential. The scanning rate was 0.5 mV/s.

3. Results

3.1. Microstructure

In Figure 2, the OM microstructure of the test samples before and after heat treatment is presented. In Figure 2a, the microstructure of the AM sample appears as a mosaic-like structure, with the minimum unit size of the mosaic structure being approximately 100×100 μm. This structure consists of larger grains in the central region and finer, fragmented grains at the edges. The unique microstructure is primarily determined by the building process, where the laser diameter is 100 μm, and the scanning interval is also 100 μm. With each layer formed at a 90° angle to the previous one, the building strategies result in the mosaic-like microstructure. Additionally, previous studies have indicated that the untreated microstructure is a unitary ferrite structure [19].

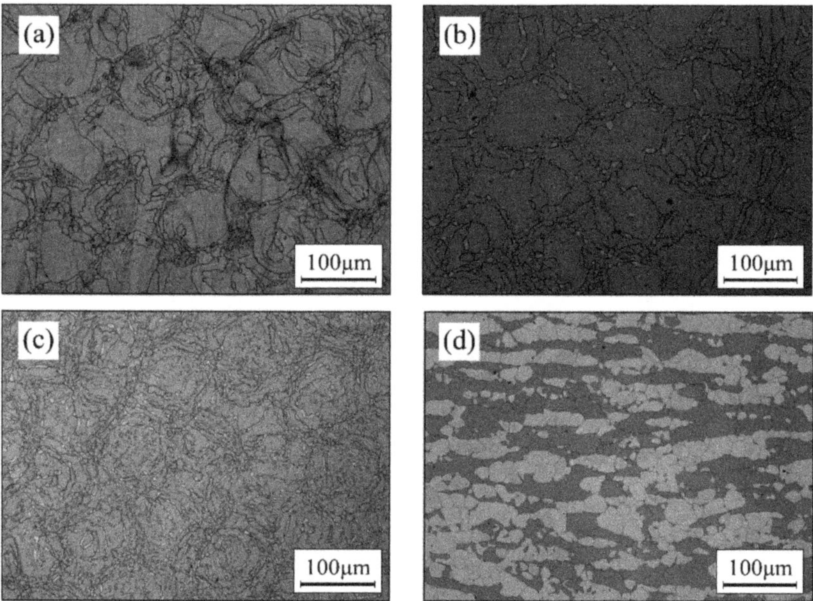

Figure 2. OM image of the specimen before and after heat treatment: (**a**) AM sample before heat treatment, (**b**) AM sample heat treated at 1200 °C for 1 h, (**c**) AM sample heat treated at 1100 °C for 1 h, and (**d**) conventionally processed sample.

Figure 2b and 2c, respectively, illustrate the OM microstructure after heat treatment at 1200 °C and 1100 °C for 1 h. Compared to the untreated microstructure, new austenite precipitated along the ferrite grain boundaries and within the grains after heat treatment, transforming the microstructure from unitary ferrite to a dual-phase structure. It is noteworthy that the regular mosaic pattern at the macroscopic level remained intact. This is mainly attributed to the low-nitrogen (N) composition design. Even after thorough heat treatment (1 h), the newly formed austenite impeded the mutual merging of the original ferrite grains.

As a reference, the microstructure of the S32750 duplex stainless steel with standard composition after conventional hot rolling and solution treatment (1100 °C $\times$ 1 h) is shown in Figure 2d. It can be observed that the microstructure of the AM samples is significantly different from that of the duplex stainless steel produced through the conventional process. The main differences lie in the phase morphology and grain size. Conventional processes including casting, forging, hot rolling, and heat treating produced larger grain sizes compared to one-off rapid prototyping of AM samples. At the same time, the conventional process 25Cr duplex stainless steel has significantly more austenite due to its higher N content. In terms of phase distribution, the two phases of AM sample were mainly distributed along the build trajectory, while the two phases of the conventional processes were mainly distributed along the rolling direction. The differences in corrosion resistance performance resulting from structural variations between the two processes are discussed in detail in the following sections.

3.2. Corrosion Resistance

The open-circuit potential (OCP) of the untreated AM sample was the lowest after solution treatment, which was slightly better than that of the conventional process. In Figure 3, OCP curves after cathodic polarization at −800 mV for 3 min are presented for several samples. After a test duration of 1800 s, the OCP values of all samples stabilized around −100 mV. The OCP value of the untreated AM sample was the most negative and was significantly lower than the other samples. After heat treatment, the OCP values noticeably increased and were all better than the conventional process samples [20]. A lower OCP value indicates the higher electrochemical activity of the tested sample. Therefore, the untreated AM sample exhibited the highest corrosion tendency and had the worst corrosion resistance, while after heat treatment, its corrosion resistance was better than that of the conventional process sample.

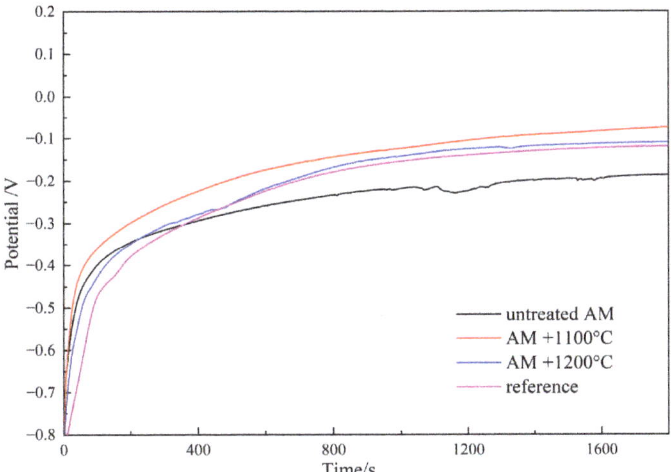

Figure 3. OCP of duplex stainless steels under different preparation processes.

Figure 4 shows the electrochemical impedance spectra of different samples at open-circuit potential. In the Nyquist diagram, all capacitance reactance arcs formed incomplete semicircles, indicating that the corrosion mechanism of the tested steel in the 3.5% NaCl solution remains unchanged regardless of the preparation method. The capacitive reactance arcs of both the untreated AM sample and the sample from conventional processes were small and relatively closed to each other. However, after heat treatment, the capacitive reactance arc of the AM sample noticeably increased, with the most significant increase observed in the sample treated at 1100 °C. This phenomenon reflects the improved corrosion resistance of the AM sample after heat treatment. As can be seen from the Bode plot, the heat-treated AM sample had a higher impedance modulus ($|Z|$) at low frequencies, indicating excellent corrosion resistance. Similarly, the phase angle plot shows excellent passivation performance.

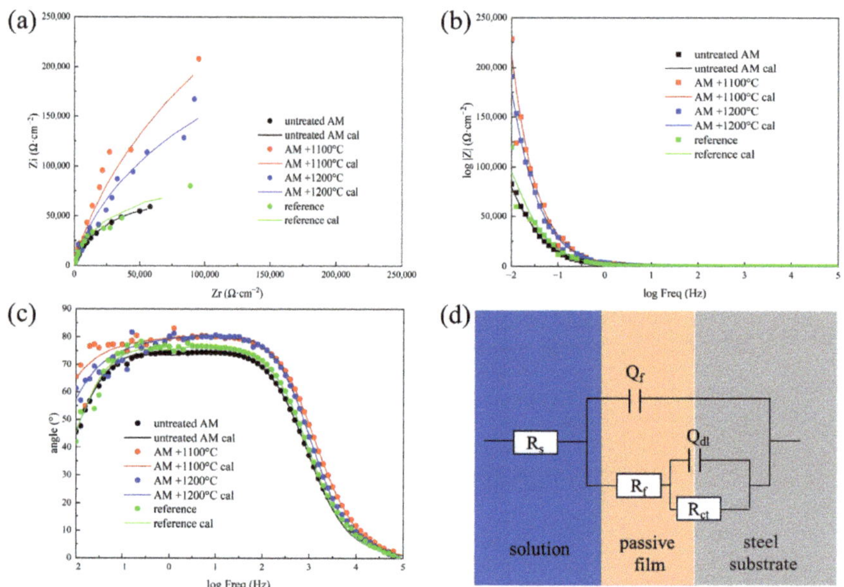

Figure 4. Impedance spectra of samples under different preparation processes: (**a**) Nyquist diagram; (**b**,**c**) Bode diagram; (**d**) equivalent circuit for fitting EIS diagram.

The impedance spectra data were fitted using the equivalent circuit shown in Figure 4d. R_s represents the electrolyte resistance, the constant phase element Q_f corresponds to the passive film capacitance, R_f is the passive film resistance, and Q_{dl} and R_{ct} represent the double-layer capacitance and charge transfer resistance, respectively [21,22]. The membrane values of different samples in the high-frequency region are closely related to the corrosion resistance of the passive film, while the low-frequency region may be associated with charge transfer resistance [20]. The fitting results using the aforementioned equivalent circuit are presented in Table 2. The combination of R_f and R_{ct} reflects the holistic corrosion resistance of the samples. From the table, it can be seen that the n values of Q_f and Q_{dl} are close to 1, indicating that the constant phase elements were close to pure capacitors. A greater value of n designates a decrease in surface inhomogeneity, associated with a strong adsorption of polymer and inhibitor [23]. This suggests effective passivation of the passive film. Comparing the values of R_f and R_{ct} for different samples, it was found that the charge transfer resistance R_{ct} was significantly larger than R_f, especially for the heat-treated AM sample. This indicates that charge transfer was more difficult, reflecting the greater difficulty for electrons to escape from the surface of the substrate, making it more challenging for metal atoms on the substrate to be oxidized into ions. The passive layer on the stainless steel surface is composed of substances such as Cr_2O_3, $Cr(OH)_3$, Fe_2O_3, Fe_3O_4, etc. [24,25]. Since there were fewer metal ions involved in the formation of these substances, the structure, thickness, and formation processes of the passive film were different from the other two samples, which is also the reason for the smaller R_f in the heat-treated AM sample. Although the resistance of electrons passing through the passive film was relatively small, the n values of Q_f for all the heat-treated AM samples were 1, indicating that the capacitive properties of the passivation film of the samples are more pronounced, and the hindering effect on the ions is more obvious.

Table 2. Fitted electrochemical parameters for EIS data of duplex stainless steels under different preparation processes.

	R_s (Ω cm^2)	Q_f (μF/cm^2)		R_f (Ωcm^2)	Q_{dl} (μF/cm^2)		R_{ct} (Ωcm^2)	$R_f + R_{ct}$ (Ωcm^2)
		Y_0	n		Y_0	n		
Untreated AM	6.53	91.97	0.855	1.82×10^4	12.25	0.978	1.27×10^5	1.46×10^5
AM + 1100 °C	5.56	13.82	1.000	7.85	34.82	0.827	9.31×10^5	9.31×10^5
AM + 1200 °C	6.16	17.95	1.000	11.12	35.7	0.805	5.22×10^5	5.22×10^5
Reference	6.06	80.08	0.874	2.76×10^4	9.959	1.000	1.41×10^5	1.68×10^5

In summary, the corrosion resistance of the test steel can be impacted by the magnitude of $R_f + R_{ct}$ value (the combination value of film resistance and charge transfer resistance). It can be seen that the corrosion resistance of the untreated AM sample was the worst, and it was significantly improved after heat treatment. The $R_f + R_{ct}$ value of the AM sample after heat treatment at 1200 °C is 3.2 times of that of the conventional processes sample, and it increased to 5.5 times at 1100 °C, which are both much better than that of the conventional sample.

Figure 5 presents the cyclic polarization curves of different samples in a 3.5% NaCl solution, showing a similar overall trend and indicating a common corrosion mechanism. Combining the curves, it is evident that the anodic curves of different samples exhibited clear passivation behavior. Through the analysis of the polarization curves, information such as the corrosion current density (I_{corr}), critical pitting potential (E_p), and corrosion potential (E_{corr}) can be obtained. When the applied current exceeded the passivation region, the current density suddenly increased, and pitting corrosion was considered to occur when the current density reached 100 μA/cm^2. The potential corresponding to this point is defined as E_p. The stability of the passive film is characterized by the difference between E_p and the corrosion potential E_{corr}. When the potential reached E_p, the curve began to reverse scan, and during the reverse scan, the current exhibited a hysteresis phenomenon. The intersection of the reverse scan curve with the forward scan anode curve is repassivation potential, defined as E_r. The difference between E_p and E_r is used to characterize the repairability of the passive film on the sample, i.e., the repassivation performance.

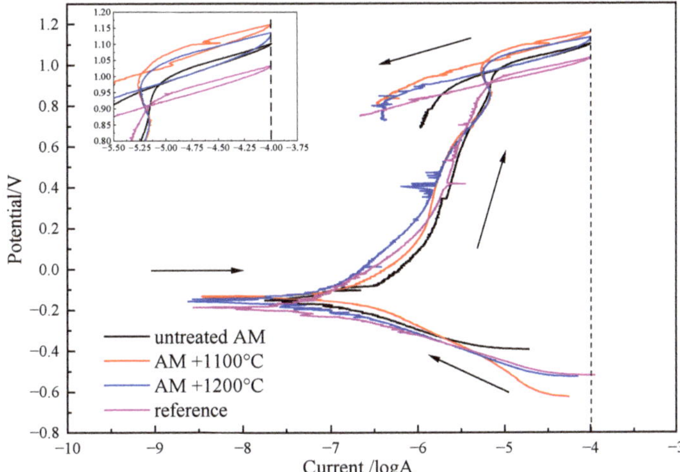

Figure 5. Potentiodynamic polarization curve of duplex stainless steels under different preparation processes.

From Table 3, it can be observed that the conventional process sample had the most negative corrosion potential, followed by the untreated AM sample. After heat treatment, the corrosion potential of the AM sample was improved, especially after the 1100 °C heat treatment, where the corrosion potential was significantly higher than other samples, indicating the least tendency for corrosion to occur. The pitting potential followed the same pattern as the corrosion potential. After heat treatment at 1100 °C and 1200 °C, the pitting potential of AM samples was 128.5 mV and 103.4 mV higher than that of conventional samples, respectively. In addition, the E_p-E_{corr} values of all AM samples were larger than that of the conventional sample, implying that the AM samples had a larger passivation interval. In contrast to the pattern of pitting potential, the conventional process sample exhibited the best repassivation performance, while the heat-treated AM samples showed relatively poor repairability of the passive film, with a larger E_p-E_r value. In summary, the AM samples after heat treatment have a lesser corrosion tendency, a higher pitting potential, and a larger passivation interval, although the repassivation performance is slightly worse; overall, the corrosion resistance is better than that of conventional process samples.

Table 3. Electrochemical parameters for polarization curve of duplex stainless steels under different preparation processes.

	E_{corr}/mV	I_{corr}/μA	E_p/mV	E_p-E_{corr}/mV	E_r/mV	E_p-E_r/mV
Untreated AM	−153.46	0.282	1100.75	1254.21	972.84	127.91
AM + 1100 °C	−134.99	0.293	1160.33	1295.32	1008.97	151.36
AM + 1200 °C	−142.91	0.265	1135.19	1278.10	985.91	149.28
Reference	−189.34	0.272	1031.80	1221.14	909.27	122.53

4. Discussion

In the above results, it is evident that the AM samples after heat treatment, despite having initially lower PREN values and higher oxygen content, exhibited better corrosion resistance performance than the conventional processed samples. This section discussed the reasons for the superior corrosion resistance performance of the AM samples and the variations in the corrosion resistance performance of AM samples under different conditions. In this section, the mechanism of corrosion resistance enhancement of AM samples is discussed in terms of phase composition, elemental distribution, substructure, and inclusions.

4.1. Influence of Phase Composition and Substructure on Corrosion Resistance Properties

For DSSs, the coexistence of two phases with a reasonable distribution of elements is conducive to the improvement of corrosion resistance performance [26]. It is generally considered that the corrosion resistance of austenite is superior to that of ferrite in dual-phase stainless steel [27]. Within the compositional range of dual-phase stainless steel, unitary ferritic structure without heat treatment is also unfavorable to corrosion resistance from an element distribution perspective.

Through OM microstructure analysis, the variation in austenite content was examined: 0% for untreated samples, 8.2% for heat-treated samples at 1200 °C, and 16.3% for heat-treated samples at 1100 °C. Under the same composition, the change in the proportion of the two phases resulting in the variation of element distribution is the main cause of the change in corrosion resistance. Figure 6 presents the EPMA mapping results of additive manufacturing samples at different heat treatment temperatures. It can be observed that after heat treatment at 1200 °C, austenite appeared in the intercrystalline regions of the AM samples, with Ni elements enriched in austenite and Cr elements enriched in ferrite [28]. After heat treatment at 1100 °C, austenite content increased, with austenite appearing not only in the intercrystalline regions but also within the grains. Ni was highly enriched in both intercrystalline and intragranular austenite. Compared to heat treatment at 1200 °C, the enrichment of Cr elements in the ferrite phase was more pronounced at this temperature.

The appearance of austenite led to a more balanced distribution of elements between the two phases, which was beneficial for improving corrosion resistance performance.

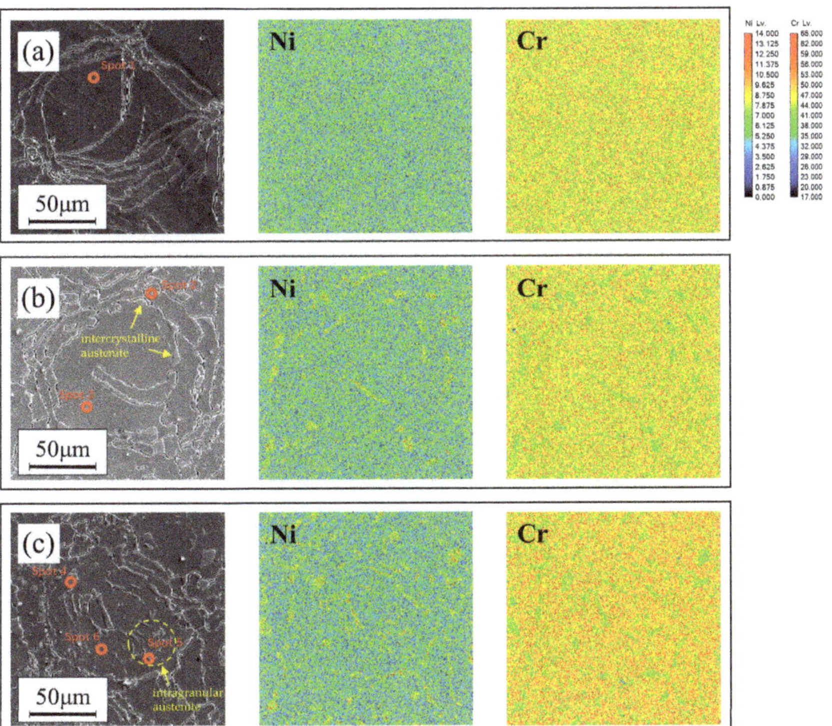

Figure 6. EPMA mapping results of AM samples of different processes: (**a**) untreated, (**b**) 1200 °C solid solution, (**c**) 1100 °C solid solution.

Furthermore, the intercrystalline regions had higher energy, providing a pathway for rapid diffusion of elements. This resulted in compositional differences between intergranular and intragranular austenite. EPMA spot scanning was conducted to characterize the elements content of different phase in AM samples under different conditions, with results shown in Table 4. For DSSs, corrosion resistance primarily depends on the content of Cr, Mo, and N elements. The N element content is mainly enriched in the austenite phase, while Cr and Mo have higher concentrations in the ferrite phase [29]. Typically, the impact of N element on corrosion resistance performance is greater than that of the other two elements. Additionally, the austenite phase exhibits a more corrosion-resistant microstructure, while ferrite often becomes the weaker phase during corrosion processes [30]. Table 4 shows that the Cr and N contents in intragranular austenite were both higher than those in intercrystalline austenite, making it the most corrosion-resistant component among intragranular austenite, intercrystalline austenite, and ferrite. Furthermore, the appearance of intragranular austenite led to further enrichment of Cr elements in the ferrite, thereby enhancing the corrosion resistance of the weakest phase (ferrite). In summary, the combination of intragranular and intercrystalline austenite phases with ferrite ensures a more rational distribution of elements, thereby improving the corrosion resistance performance.

For AM samples, the difference in corrosion resistance between the samples before and after heat treatment is not only due to the ratio of the two phases, the morphology of the two phases, and the elemental distribution between the two phases but is also affected by the number of grains and substructures [31–33]. During the building forming process, the central part of the laser had higher energy, allowing grains to grow sufficiently, while the edges, due to the reduction in laser energy, have relatively smaller grain sizes. Smaller grain size implies more grain boundaries, substructures, and dislocation density.

Figure 7 illustrates the EBSD kernel average misorientation (KAM) test results for AM specimens in different states. KAM can be used to characterize the stress state and deformation extent [11,34]. None of the AM samples used for testing in this work were plastically deformed, so a higher KAM indicates a higher stress state. Therefore, it is inferred that the high KAM originates from the high dislocation density. The sources of these dislocations were mainly thermal stress residues and supersaturated solid solutions of elements during rapid cooling. After heat treatment, thermal stresses were removed, elements were redistributed between the two phases, the peak of the KAM value was shifted to the left, the peak width decreased, and the dislocation density decreased.

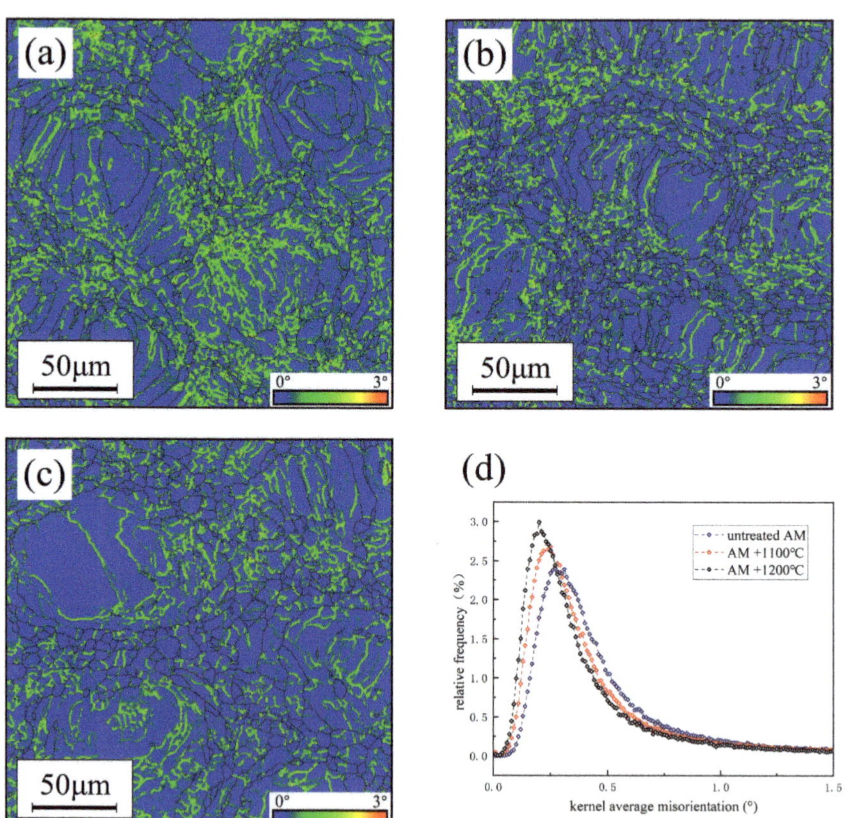

Figure 7. EBSD KAM test results of AM sample: (**a**) untreated, (**b**) 1200 °C solid solution, (**c**) 1100 °C solid solution, and (**d**) KAM distribution.

The untreated AM samples exhibited a unitary ferritic structure. However, at room temperature, the saturation N element solubility in ferrite is only 0.07% [35], while the N content of the AM sample was 0.09%, which was entirely oversaturated and dissolved in the ferrite. Figure 8 shows TEM images of the untreated samples, revealing that even without deformation, the ferrite has an extremely high dislocation density due to the oversaturation of N elements and residual thermal stress from the rapid cooling during the forming process. This is one of the reasons why the corrosion resistance performance of untreated samples was lower than that of heat-treated samples.

Table 4. EPMA spot scanning results.

Spot	Phase	The Element Content (wt.%)						
		Ni	Cr	Mo	Mn	C	N	Fe
1	Ferrite	6.63	25.30	3.65	0.48	0.64	0.15	Bal.
2	Intercrystalline austenite	8.87	22.33	2.74	0.41	0.43	0.61	Bal.
3	Ferrite	6.32	25.69	3.67	0.43	0.60	0.05	Bal.
4	Intercrystalline austenite	8.33	22.59	2.81	0.64	0.54	0.36	Bal.
5	Intragranular austenite	7.00	24.65	3.25	0.75	0.51	0.41	Bal.
6	Ferrite	6.07	25.99	3.72	0.52	0.59	0.06	Bal.

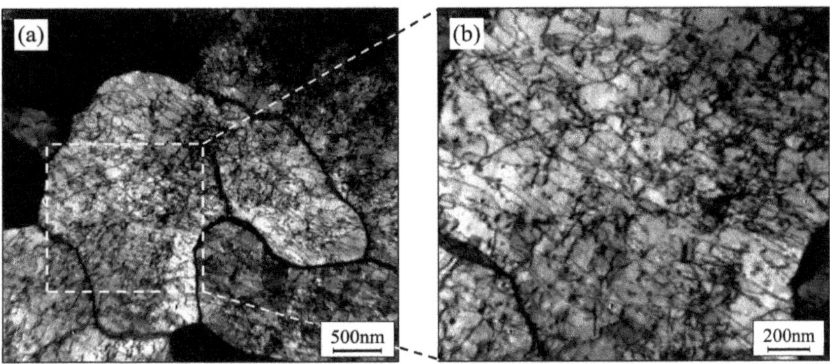

Figure 8. TEM morphology of dislocations in untreated AM sample (**a**) and selected area enlargement (**b**).

4.2. Effect of Inclusions on Corrosion Resistance

There are various factors influencing the pitting corrosion resistance of stainless steel. For conventional process stainless steel, inclusions are one of the significant contributors to the deterioration of its corrosion resistance. During the conventional smelting process, unavoidable steps like deoxidation introduce inclusions such as Al_2O_3, MnO, and others [36,37]. These inclusions have higher melting points than the alloy, and their stable nature makes it challenging to eliminate them through subsequent processing steps. These inclusions are not only difficult to eliminate but also have relatively large sizes, leading to a certain degradation in the final mechanical and corrosion resistance properties of stainless steel. Therefore, the modification of inclusions and the enhancement of the purity of stainless steel are crucial measures for optimizing its corrosion resistance performance.

Figure 9a,b, respectively, show surface micrographs of the polished states of the AM samples and the conventionally processed samples. It can be observed that the AM samples exhibited extremely high purity, with virtually no visible inclusions. In contrast, the conventionally processed sample showed a significant presence of black inclusions. Furthermore, the distribution of inclusions was statistically analyzed using ASPEX inclusion analysis, with the results depicted in Figure 9c,d. It should be noted that the substances detected by ASPEX are still customarily referred to as inclusions, even though they are much smaller than common inclusions and are thus unobservable under a metallurgical microscope [14]. In the untreated AM samples, a minimal number of inclusions was observed. Conversely, the conventional process samples exhibited a significantly higher number of inclusions, which became one of the significant factors deteriorating its corrosion resistance performance. This is why, despite having a higher PREN value, the corrosion resistance performance of the conventional processed samples was weaker than that of the AM samples.

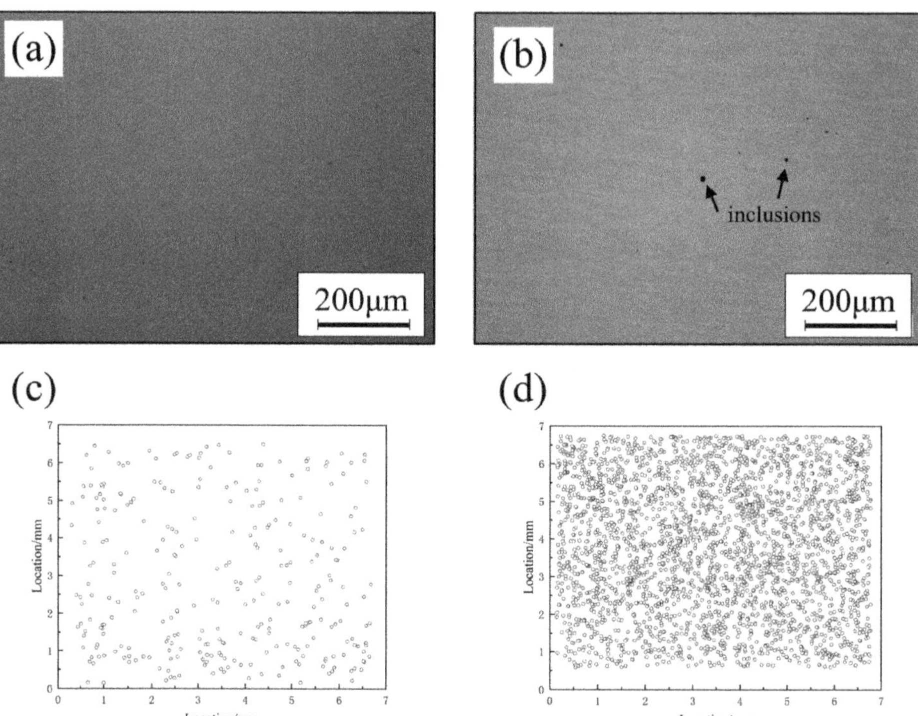

Figure 9. Distribution of inclusions: (**a**) AM sample surface morphology (OM), (**b**) conventionally processed surface morphology (OM), (**c**) AM sample inclusion distribution (ASPEX), and (**d**) conventional processed sample inclusion distribution (ASPEX).

However, it is worth noting that the oxygen (O) element content in the AM samples was as high as 0.02%, much higher than the oxygen content in the conventionally processed samples (0.002%). Interestingly, O, as the primary forming element of inclusions, did not appear in the form of large-sized inclusions in the AM samples. To investigate this, we explored the presence of oxygen in the AM samples. Figure 10 presents the TEM microstructure of the AM samples, revealing numerous particles with sizes ranging from 20 to 100 nm. STEM characterization of their elemental composition indicated the presence of particles rich in oxygen, manganese, and chromium. The formation of these nanoscale oxides was primarily attributed to the unique process of AM.

Through the powder bed fusion process, high-purity powder without inclusions could be obtained. However, the powder-making process unavoidably introduced oxygen, leading to a higher oxygen content. In the additive manufacturing process, the laser created a high-temperature melt pool on the powder bed, where numerous oxides rapidly formed. Due to the relatively fast scanning speed (850 mm/s) and the short duration of the melt pool, coupled with the extremely rapid cooling rate (10^5–10^7 K/s) [35], these oxides did not have enough time to grow into large inclusions. Instead, they formed nanoscale inclusions. The nanoscale inclusions significantly enhanced the macroscopic purity compared to conventional processes, which ultimately resulted in better corrosion resistance performance at a lower PREN in the AM samples.

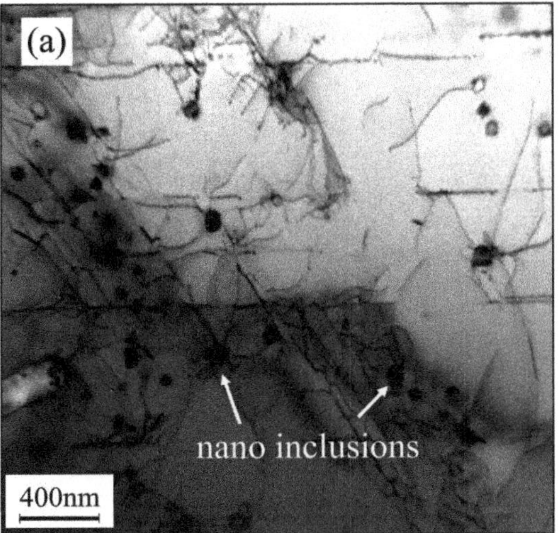

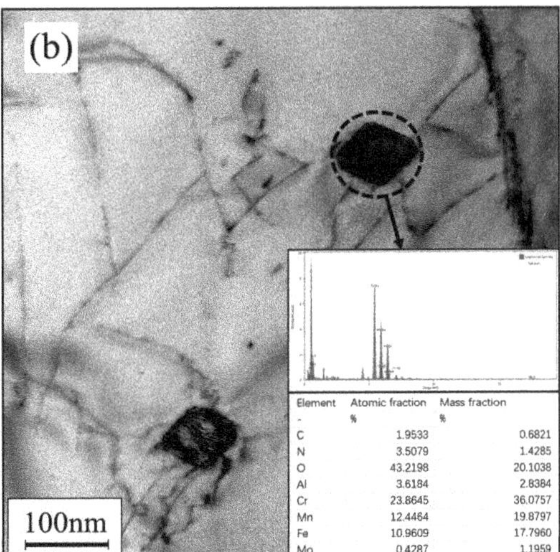

Element	Atomic fraction %	Mass fraction %
C	1.9533	0.6821
N	3.5079	1.4285
O	43.2198	20.1038
Al	3.6184	2.8384
Cr	23.8645	36.0757
Mn	12.4464	19.8797
Fe	10.9609	17.7960
Mo	0.4287	1.1959

Figure 10. TEM morphology of inclusions in heat-treated AM samples: (**a**) 25,000× magnification; (**b**) 100,000× magnification.

5. Conclusions

A 0.09% N 25Cr-type duplex stainless steel was prepared by AM and heat treatment, and its corrosion resistance was investigated. The results indicate that the refinement of inclusion size has a favorable impact on its corrosion resistance performance. With a lower PREN value, additive manufacturing products can achieve superior actual corrosion resistance performance.

(1) The corrosion resistance of 0.09% N 25Cr-type AM test steel after heat treatment is superior to that of S32750 duplex stainless steel with the same Cr content when hot rolled and in solid solution condition. In a 3.5% wt.% NaCl solution, the combination value of film resistance and charge transfer resistance of AM samples in solid solution at 1100 °C and 1200 °C was 5.5 times and 3.2 times that of conventional S32750 duplex stainless steel, respectively; the pitting potential was 128.5 mV and 103.4 mV higher, respectively;

(2) The disappearance of residual thermal stress and the reasonable distribution of Cr and N elements in the two phases are the reasons for the improvement of the corrosion resistance of AM duplex stainless steel after heat treatment. When the heat treatment temperature decreased from 1200 °C to 1100 °C, the austenite phase morphology changed from intercrystalline austenite to intercrystalline and intragranular austenite. The Cr and N contents in intragranular austenite were both higher than those in the intercrystalline austenite, also leading to further enrichment of Cr elements in the ferrite and enhancing the corrosion resistance of AM DSSs;

(3) The influence of inclusions on corrosion resistance of DSSs is more obvious than PREN value. Unlike the large particle inclusions commonly found in duplex stainless steel prepared by conventional processes, the oxygen in AM samples only exists in nanoscale Mn and Cr oxides due to the particularity of AM technology. The extremely high purity of AM duplex stainless steel with no visible inclusions resulted in a higher corrosion resistance exhibited at lower PREN values.

Author Contributions: Conceptualization, J.H., Z.S. and H.F.; Methodology, Y.G. and C.W.; Validation, J.H.; Investigation, Y.G. and J.L.; Resources, C.W.; Data curation, X.W.; Writing—original draft, Y.G. and J.L.; Writing—review & editing, Y.G. and J.H.; Supervision, Z.S. and H.F.; Project administration, J.H. All authors have read and agreed to the published version of the manuscript.

Funding: This research received no external funding.

Institutional Review Board Statement: Not applicable.

Informed Consent Statement: Not applicable.

Data Availability Statement: Data are contained within the article.

Conflicts of Interest: All authors were employed by the company Research Institute of Special Steels, Central Iron & Steel Research Institute Co., Ltd. The authors declare that the research was conducted in the absence of any commercial or financial relationships that could be construed as a potential conflict of interest.

References

1. Freitas, B.J.M.; Rodrigues, L.C.M.; Claros, C.A.E.; Botta, W.J.; Koga, G.Y.; Bolfarini, C. Ferritic-Induced High-Alloyed Stainless Steel Produced by Laser Powder Bed Fusion (L-PBF) of 2205 Duplex Stainless Steel: Role of Microstructure, Corrosion, and Wear Resistance. *J. Alloy. Compd.* **2022**, *918*, 165576. [CrossRef]
2. Ha, H.; Kang, J.; Lee, T.; Park, H.; Kim, H. Effect of Grain Size on the Pitting Corrosion Resistance of Lean Duplex Stainless Steel. *Steel Res. Int.* **2023**, *94*, 2200227. [CrossRef]
3. Mondal, R.; Bonagani, S.K.; Raut, P.; Kumar, S.; Sivaprasad, P.V.; Chai, G.; Kain, V.; Samajdar, I. Dynamic Recrystallization and Phase-Specific Corrosion Performance in a Super Duplex Stainless Steel. *J. Mater. Eng. Perform.* **2022**, *31*, 1478–1492. [CrossRef]
4. Wu, X.; Song, Z.; Liu, L.; He, J.; Zheng, L. Effect of Secondary Austenite on Fatigue Behavior of S32750 Super Duplex Stainless Steel. *Mater. Lett.* **2022**, *322*, 132487. [CrossRef]
5. Yang, Y.; Pan, X. Effect of Mn/N Ratio on Microstructure and Mechanical Behavior of Simulated Welding Heat Affected Zone in 22% Cr Lean Duplex Stainless Steel. *Mater. Sci. Eng. A* **2022**, *835*, 142676. [CrossRef]
6. Zhi-gang, S.H.F.S. Development of Chinese Duplex Stainless Steel in Recent Years. *J. Iron Steel Res. Int.* **2017**, *24*, 121–130.
7. Chen, M.; He, J.; Li, J.; Liu, H.; Xing, S.; Wang, G. Excellent Cryogenic Strength-Ductility Synergy in Duplex Stainless Steel with Heterogeneous Lamella Structure. *Mater. Sci. Eng. A* **2022**, *831*, 142335. [CrossRef]
8. Abedi, H.R.; Hanzaki, A.Z.; Azami, M.; Kahnooji, M.; Rahmatabadi, D. The High Temperature Flow Behavior of Additively Manufactured Inconel 625 Superalloy. *Mater. Res. Express* **2019**, *6*, 116514. [CrossRef]
9. Jiang, D.; Birbilis, N.; Hutchinson, C.R.; Brameld, M. On the Microstructure and Electrochemical Properties of Additively Manufactured Duplex Stainless Steels Produced Using Laser-Powder Bed Fusion. *Corrosion* **2020**, *76*, 871–883. [CrossRef]
10. Núñez De La Rosa, Y.E.; Palma Calabokis, O.; Borges, P.C.; Ballesteros Ballesteros, V. Effect of Low-Temperature Plasma Nitriding on Corrosion and Surface Properties of Duplex Stainless Steel UNS S32205. *J. Mater. Eng. Perform.* **2020**, *29*, 2612–2622. [CrossRef]
11. Iams, A.D.; Keist, J.S.; Palmer, T.A. Formation of Austenite in Additively Manufactured and Post-Processed Duplex Stainless Steel Alloys. *Metall. Mater. Trans. A Phys. Metall. Mater. Sci.* **2020**, *51*, 982–999. [CrossRef]
12. Yan, F.; Xiong, W.; Faierson, E.; Olson, G.B. Characterization of Nano-Scale Oxides in Austenitic Stainless Steel Processed by Powder Bed Fusion. *Scr. Mater.* **2018**, *155*, 104–108. [CrossRef]
13. Zhang, J.; Dong, H.; Xi, X.; Tang, H.; Li, X.; Rao, J.H.; Xiao, Z. Enhanced Mechanical Performance of Duplex Stainless Steels Via Dense Core-Shell Nano-Inclusions In-Situ Formed upon Selective Laser Melting. *Scr. Mater.* **2023**, *237*, 115711. [CrossRef]
14. Haghdadi, N.; Laleh, M.; Chen, H.; Chen, Z.; Ledermueller, C.; Liao, X.; Ringer, S.; Primig, S. On the Pitting Corrosion of 2205 Duplex Stainless Steel Produced by Laser Powder Bed Fusion Additive Manufacturing in the As-Built and Post-Processed Conditions. *Mater. Des.* **2021**, *212*, 110260. [CrossRef]
15. Laleh, M.; Haghdadi, N.; Hughes, A.E.; Primig, S.; Tan, M.Y.J. Enhancing the Repassivation Ability and Localised Corrosion Resistance of an Additively Manufactured Duplex Stainless Steel by Post-Processing Heat Treatment. *Corros. Sci.* **2022**, *198*, 110106. [CrossRef]
16. Gor, M.; Soni, H.; Wankhede, V.; Sahlot, P.; Grzelak, K.; Szachgluchowicz, I.; Kluczyński, J. A Critical Review on Effect of Process Parameters on Mechanical and Microstructural Properties of Powder-Bed Fusion Additive Manufacturing of SS316L. *Materials* **2021**, *14*, 6527. [CrossRef] [PubMed]
17. Zhang, D.; Liu, A.; Yin, B.; Wen, P. Additive Manufacturing of Duplex Stainless Steels—A Critical Review. *J. Manuf. Process.* **2022**, *73*, 496–517. [CrossRef]
18. Zhang, Y.; Cheng, F.; Wu, S. Improvement of Pitting Corrosion Resistance of Wire Arc Additive Manufactured Duplex Stainless Steel through Post-Manufacturing Heat-Treatment. *Mater. Charact.* **2021**, *171*, 110743. [CrossRef]
19. He, J.; Lv, J.; Song, Z.; Wang, C.; Feng, H.; Wu, X.; Zhu, Y.; Zheng, W. Maintaining Excellent Mechanical Properties via Additive Manufacturing of Low-N 25Cr-Type Duplex Stainless Steel. *Materials* **2023**, *16*, 7125. [CrossRef]
20. Zhu, M.; Zhang, Q.; Yuan, Y.; Guo, S. Effect of Microstructure and Passive Film on Corrosion Resistance of 2507 Super Duplex Stainless Steel Prepared by Different Cooling Methods in Simulated Marine Environment. *Int. J. Miner. Metall. Mater.* **2020**, *27*, 1100–1114. [CrossRef]
21. Kocijan, A.; Merl, D.K.; Jenko, M. The Corrosion Behaviour of Austenitic and Duplex Stainless Steels in Artificial Saliva with the Addition of Fluoride. *Corros. Sci.* **2011**, *53*, 776–783. [CrossRef]

22. Ming, J.; Shi, J. Chloride Resistance of Cr-bearing Alloy Steels in Carbonated Concrete Pore Solutions. *Int. J. Miner. Metall. Mater.* **2020**, *27*, 494–504. [CrossRef]
23. Moshtaghi, M.; Eškinja, M.; Mori, G.; Griesser, T.; Safyari, M.; Cole, I. The Effect of HPAM Polymer for Enhanced Oil Recovery on Corrosion Behaviour of a Carbon Steel and Interaction with the Inhibitor Under Simulated Brine Conditions. *Corros. Sci.* **2023**, *217*, 111118. [CrossRef]
24. Li, B.; Qu, H.; Lang, Y.; Feng, H.; Chen, Q.; Chen, H. Copper Alloying Content Effect on Pitting Resistance of Modified 00Cr20Ni18Mo6CuN Super Austenitic Stainless Steels. *Corros. Sci.* **2020**, *173*, 108791. [CrossRef]
25. Zhang, J.; Zhu, Y.; Xi, X.; Xiao, Z. Altered Microstructure Characteristics and Enhanced Corrosion Resistance of UNS S32750 Duplex Stainless Steel Via Ultrasonic Surface Rolling Process. *J. Mater. Process. Tech.* **2022**, *309*, 117750. [CrossRef]
26. Han, Y.; Liu, Z.; Wu, C.; Zhao, Y.; Zu, G.; Zhu, W.; Ran, X. A Short Review on the Role of Alloying Elements in Duplex Stainless Steels. *Tungsten* **2022**, *5*, 419–439. [CrossRef]
27. Zhang, Y.; Wu, S.; Cheng, F. A Specially-Designed Super Duplex Stainless Steel with Balanced Ferrite: Austenite Ratio Fabricated Via Flux-Cored Wire Arc Additive Manufacturing: Microstructure Evolution, Mechanical Properties and Corrosion Resistance. *Mater. Sci. Eng. A* **2022**, *854*, 143809. [CrossRef]
28. Biserova-Tahchieva, A.; Biezma-Moraleda, M.V.; Llorca-Isern, N.; Gonzalez-Lavin, J.; Linhardt, P. Additive Manufacturing Processes in Selected Corrosion Resistant Materials: A State of Knowledge Review. *Materials* **2023**, *16*, 1893. [CrossRef] [PubMed]
29. Ko, G.; Kim, W.; Kwon, K.; Lee, T. The Corrosion of Stainless Steel Made by Additive Manufacturing: A Review. *Metals* **2021**, *11*, 516. [CrossRef]
30. Tavares, S.S.M.; Gonzaga, A.C.; Pardal, J.M.; Conceição, J.N.; Correa, E.O. Nitrides Precipitation and Preferential Pitting Corrosion of Ferrite Phase in UNS S39274 Superduplex Stainless Steel. *Metallogr. Microstruct. Anal.* **2020**, *9*, 685–694. [CrossRef]
31. David, C.; Ruel, F.; Krajcarz, F.; Boissy, C.; Saedlou, S.; Vignal, V. Effect of Grain Size on the Anodic Dissolution of Lean Duplex UNS S32202 Austenitic-Ferritic Stainless Steel. *Corrosion* **2019**, *75*, 1450–1460. [CrossRef] [PubMed]
32. Koga, N.; Noguchi, M.; Watanabe, C. Low-Temperature Tensile Properties, Deformation and Fracture Behaviors in the Ferrite and Austenite Duplex Stainless Steel with Various Grain Sizes. *Mater. Sci. Eng. A* **2023**, *880*, 145354. [CrossRef]
33. Mondal, R.; Rajagopal, A.; Bonagani, S.K.; Prakash, A.; Fuloria, D.; Sivaprasad, P.V.; Chai, G.; Kain, V.; Samajdar, I. Solution Annealing of Super Duplex Stainless Steel: Correlating Corrosion Performance with Grain Size and Phase-Specific Chemistry. *Metall. Mater. Trans. A Phys. Metall. Mater. Sci.* **2020**, *51*, 2480–2494. [CrossRef]
34. Wroński, S.; Tarasiuk, J.; Bacroix, B.; Baczmański, A.; Braham, C. Investigation of Plastic Deformation Heterogeneities in Duplex Steel by EBSD. *Mater. Charact.* **2012**, *73*, 52–60. [CrossRef]
35. Saeidi, K.; Kevetkova, L.; Lofaj, F.; Shen, Z. Novel Ferritic Stainless Steel Formed by Laser Melting from Duplex Stainless Steel Powder with Advanced Mechanical Properties and High Ductility. *Mater. Sci. Eng. A* **2016**, *665*, 59–65. [CrossRef]
36. Gao, J.; Jiang, Y.; Deng, B.; Ge, Z.; Li, J. Determination of Pitting Initiation of Duplex Stainless Steel Using Potentiostatic Pulse Technique. *Electrochim. Acta* **2010**, *55*, 4837–4844. [CrossRef]
37. Garfias, L.F.; Siconolfi, D.J. In Situ High-Resolution Microscopy on Duplex Stainless Steels. *J. Electrochem. Soc.* **2000**, *147*, 2525–2531. [CrossRef]

 *materials*

Article

Strengthening Mechanism and Damping Properties of SiC$_f$/Al-Mg Composites Prepared by Combining Colloidal Dispersion with a Squeeze Melt Infiltration Process

Guanzhang Lin [1,*], Jianjun Sha [1,2,*], Yufei Zu [1], Jixiang Dai [1], Cheng Su [1] and Zhaozhao Lv [3]

1   Key Laboratory of Advanced Technology for Aerospace Vehicles of Liaoning Province, Dalian University of Technology, Dalian 116024, China; yfzu@dlut.edu.cn (Y.Z.); jxdai@dlut.edu.cn (J.D.); sucheng@mail.dlut.edu.cn (C.S.)
2   State Key Laboratory of Structural Analysis, Optimization and CAE Software for Industrial Equipment, Dalian University of Technology, Dalian 116024, China
3   School of Defence Science & Technology, Xi'an Technological University, Xi'an 710032, China; lvzhaozhao@xatu.edu.cn
*   Correspondence: linguanzhang@mail.dlut.edu.cn (G.L.); jjsha@dlut.edu.cn (J.S.)

Abstract: SiC-fiber-reinforced Al-Mg matrix composites with different mass fractions of Mg were fabricated by combining colloidal dispersion with a squeeze melt infiltration process. The microstructure, mechanical and damping properties, and the corresponding mechanisms were investigated. Microstructure analyses found that SiC$_f$/Al-Mg composites presented a homogeneous distribution of SiC fibers, and the relative density was higher than 97% when the mass fraction of Mg was less than 20%; the fiber–matrix interface bonded well, and no obvious reaction occurred at the interface. The SiC$_f$/Al-10Mg composite exhibited the best flexural strength (372 MPa) and elastic modulus (161.7 GPa). The fracture strain of the composites decreased with an increase in the mass fraction of Mg. This could be attributed to the strengthened interfacial bonding due to the introduction of Mg. The damping capacity at RT increased dramatically with an increase in the strain when the strain amplitude was higher than 0.001%, which was better than the alloys with similar composition, demonstrating a positive effect of the SiC fiber on improving the damping capacity of composite; the damping capacity at a temperature beyond 200 °C indicated a monotonic increase tendency with the testing temperature. This could be attributed to the second phase, which formed more strong pinning points and increased the dislocation energy needed to break away from the strong pinning points.

Keywords: silicon carbide fiber; Al-based matrix composite; strengthening mechanism; mechanical properties; damping properties

Citation: Lin, G.; Sha, J.; Zu, Y.; Dai, J.; Su, C.; Lv, Z. Strengthening Mechanism and Damping Properties of SiC$_f$/Al-Mg Composites Prepared by Combining Colloidal Dispersion with a Squeeze Melt Infiltration Process. *Materials* **2024**, *17*, 1600. https://doi.org/10.3390/ma17071600

Academic Editors: Lehua Liu, Haokai Dong, Haizhou Lu and Chao Zhao

Received: 3 March 2024
Revised: 26 March 2024
Accepted: 28 March 2024
Published: 31 March 2024

1. Introduction

Aluminum-based matrix composites (AMCs) have a wide range of uses in the aerospace and transportation industries because of their high specific strength and stiffness, low coefficient of thermal expansion (CTE), and good resistance to corrosion [1,2]. Among different reinforcement approaches, carbon and silicon carbide (SiC) fibers have high priority to be used as reinforcements for AMCs due to their low density, large aspect ratio, excellent mechanical properties, low CTE, and high thermal and chemical stabilities [3,4]. Although carbon fiber is less expensive for AMC manufacturing and has a lower density than SiC fiber [5,6], the challenge is that the aluminum matrix cannot completely fill the carbon fiber bundles due to the poor wettability of the carbon fibers when combined with the aluminum [7]. Additionally, harmful interfacial reactions between carbon fibers and aluminum melt may take place during the fabrication process, which would damage the carbon fibers and lead to strong fiber–matrix interface bonding [8,9], resulting in low mechanical properties. As a result, SiC fibers are emerging as a better reinforcement solution for AMCs. Furthermore, the limited damping capacity of AMCs has restricted their use in some vibration-sensitive areas, such as in the structural components for space cameras, which require materials with favorable damping properties [10–12]. On the other hand, different mechanisms are responsible for the damping and strengthening of AMCs. Thus, it is essential to improve the mechanical properties and damping properties of AMCs simultaneously.

In recent years, great efforts have been made regarding the damping behavior of AMCs and various methods have been proposed to improve the mechanical properties and damping properties of AMCs [13–17]. Theo et al. [13] introduced different volume fractions of martensitic stainless steel and silicon carbide particles into Al-based matrix composites. They found that two-particle co-reinforced Al-Zn composites offered better damping and mechanical properties than those of SiC-particle-reinforced ones. Ram et al. [14] prepared an aluminum matrix composite with randomly distributed carbon fibers using the high-pressure infiltration method and investigated the effect of carbon fiber content on the damping capacity. The results showed that the peak damping of the composite firstly increased and then decreased with the increase in carbon fiber content, while the off-peak damping always increased with the increase in carbon fiber content. This phenomenon is believed to be related to the micro plasticity of the aluminum matrix and the dislocation breakaway at the interface.

As we know, the matrix is an important component for a composite, and its composition has a substantial impact on the properties of composites. Xu [18] selected the SiC fiber and used the vacuum pressure infiltration method to prepare continuous SiC-fiber-reinforced aluminum composites, where the volume fraction of the SiC fiber was designed to be 40% and different types of alloys (ZL102, ZL114A, ZL205A and ZL301) were used as the matrix. The effects of different matrix alloys on the interface structure, fiber damage, and fracture behavior were investigated. Chu et al. [19] investigated the influence of matrix categories on the damping capacity of SiC-fiber-reinforced aluminum matrix composites. The results revealed that the dominant damping mechanisms for SiC_f/Al composites were dislocation damping at low temperatures (<150 °C) and grain boundary damping and interface damping at high temperatures (>150 °C). Hence, the microstructure of the matrix, such as the dislocation, grain size, the angle of the grain boundary, and the interface between the matrix and the fiber, essentially affected the damping capacity and dynamic modulus of composites across the whole temperature range. This work indicated that the damping capacity and mechanical strength in the SiC_f/Al composites are not absolutely in conflict with each other, and can be achieved by adjusting the matrix composition. In terms of selecting the alloy composition of the matrix, it is evident that the introduction of Mg can not only improve the wettability of the SiC_f and Al matrix [20], but also strengthen the matrix. Li Z et al. found that the increase in Mg content was beneficial for the improvement of the damping properties of the alloy at room temperature, to a certain extent [21]. Therefore, the introduction of Mg in the matrix may have a positive effect on the improvement of both of mechanical properties and damping properties of SiC_f/Al composites.

Therefore, it is of practical significance to study the effect of Mg content on the comprehensive properties of SiC_f/Al-Mg composites. In this work, the SiC fiber and Al matrix doping with different Mg content were used to prepare the SiC_f/Al-Mg composites. The SiC_f/Al-Mg composites were prepared by combining the colloidal dispersion with the squeeze melt infiltration technique. The effects of Mg content on the microstructure, mechanical properties, and damping properties of SiC_f/Al-Mg composites were studied.

2. Materials and Methods

2.1. Raw Materials

Al powders (average size: 50 μm, purity: >99.5%, supplied by Damao Co., Ltd., Tianjin, China) and Mg powders (average size: 100 μm, purity: >99.5%, supplied by Damao Co., Ltd., Tianjin, China) were used as the raw materials. SiC fibers (Cansas3203, supplied by

Liyaxincai Co., Ltd., Quanzhou, China) with a length of about 500 μm and a diameter of 12 μm were used as the reinforcement to prepare SiC$_f$/Al composites. To regulate the organization and properties of the SiC$_f$/Al-Mg composites, Mg elements with mass fractions of 5%, 10%, 15%, and 20% were added.

2.2. Fabrication of SiC$_f$/Al-Mg Composites

Figure 1 shows the schematic diagram of the preparation process of SiC$_f$/Al-Mg composites. Firstly, the SiC fibers were debonded by pre-treatment at 400 °C in a muffle furnace under atmospheric conditions, and then the debonded fibers were cut into short SiC fibers. Subsequently, a certain amount of hydroxyethyl cellulose (HEC) was dissolved in deionized water to form an HEC colloidal solution. Then, the short SiC fibers were poured into the colloidal solution and stirred for 20 min to fully disperse them in the HEC colloidal solution. The aluminum and magnesium powders were gradually poured into the colloidal solution dispersed with SiC$_f$ and stirring was continued to make them fully dispersed.

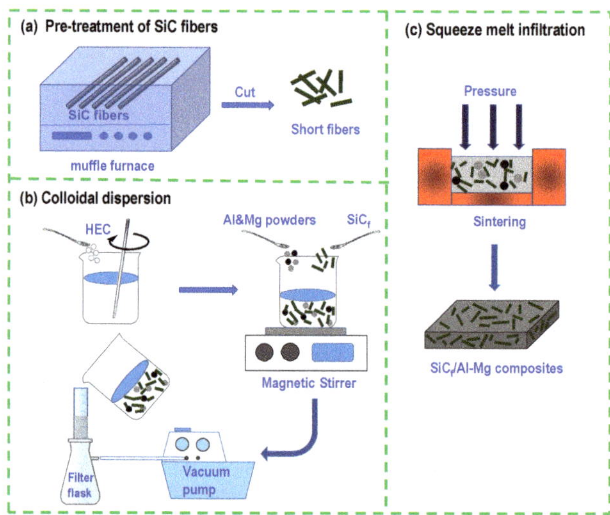

Figure 1. Schematic diagram of the preparation process of SiC$_f$/Al-Mg composites.

The excess liquid was then extracted under negative pressure using a suction–filtration device to obtain a homogeneous mixture of SiC$_f$, Al, and Mg powders. The mixture was moved into the mold and transferred into a vacuum-sintering furnace. When the furnace was heated to 680 °C, a pressure of 40 MPa was applied, and maintained for 15 min. Afterwards, the furnace was naturally cooled down to RT. Finally, a solidified block composite was obtained. The nominal volume fraction of SiC$_f$ in the composite was 20%. The mass fractions of Mg in the Al matrix were 5%, 10%, 15%, and 20%, respectively. Correspondingly, they were labelled SiC$_f$/Al-5Mg, SiC$_f$/Al-10Mg, SiC$_f$/Al-15Mg, and SiC$_f$/Al-20Mg. In a control experiment, the same process was applied to prepare the SiC$_f$-reinforced pure Al matrix composite labelled SiC$_f$/Al.

2.3. Characterization

The density of the composites was measured using Archimedes' method. The consisting phases were characterized by X-ray diffraction (XRD) (D/Max 2400, Rigaku Co., Tokyo, Japan). The Vickers hardness of the composites was tested on a Vickers hardness tester at RT with a load of 5 Kg for 10 s. The morphologies of the polished and fractured surfaces of the composites were analyzed using FE-SEM (NOVA NanoSEM 450, FEI, Hillsboro, OR,

USA) equipped with an Energy Dispersive Spectrum (EDS). The strength was carried out on the test bars with $25 \times 2.5 \times 2$ mm^3 (length by width by thickness, respectively) using a 3-point flexural test at RT. At least three bars were used for each composite. Each test was loaded with a crosshead speed of 1 mm/min. The elastic modulus was obtained based on the strain measured by the strain gauge sensor.

The damping capacity of the composites were measured with single cantilever mode according to the standard test method: ASTM E756-05. The composites were machined into samples with dimensions of $1 \times 5 \times 30$ mm^3. The temperature dependence of the damping capacity was tested by dynamic mechanical analysis (DMA) (DMA-Q800, TA, New Castle, DE, USA) at temperatures ranging from RT to 400 °C with a heating rate of 5 °C/min, under a strain amplitude of 3×10^{-4} and a frequency of 1 Hz. The strain dependence of the damping capacity was also measured by DMA, with strains ranging from 1×10^{-5} to 2.5×10^{-4} at RT with a frequency of 1 Hz.

3. Results and Discussion

3.1. Microstructure

Figure 2 shows the polished surface morphologies of composites. When the mass fraction of Mg was less than 20%, it was found that the composites had a dense structure and a homogeneous distribution of SiC fibers, and none of them showed evidence of SiC fiber aggregation (Figure 2a–c). This is because fibers in an aluminum matrix can be successfully dispersed via colloidal dispersion [22]. Figure 2d shows the morphology of the SiC$_f$/Al-20Mg composite. As indicated by the arrows in the image, it was apparent that the composite contained pores. The pores in the SiC$_f$/Al-20Mg composite could be attributed to the high Mg content. Furthermore, it was clear that the SiC fibers were bonded to the aluminum matrix and did not exhibit the presence of a reactive phase at the interface.

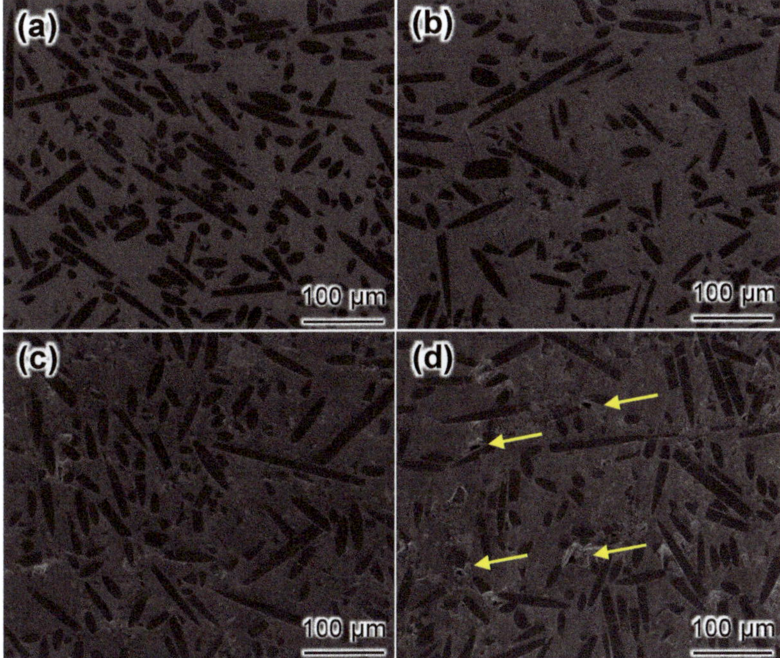

Figure 2. Morphologies of composites: (**a**) SiC$_f$/Al-5Mg; (**b**) SiC$_f$/Al-10Mg; (**c**) SiC$_f$/Al-15Mg; (**d**) SiC$_f$/Al-20Mg. Arrows in (**d**) denote pores located at surface.

According to the aluminum–magnesium phase diagram, if the magnesium content in the aluminum matrix is high, the melting point of the alloy will be lowered, which results in the composites being unable to maintain sufficient pressure during solidification [20]. As a result, pores would occur in the composites. Additionally, the microscopic morphologies of the four composites revealed that SiC fibers exhibited good isotropy and no obvious orientation distribution, indicating that the current process produced composites with good isotropy.

In order to understand the relationship between the composites' density and the Mg content clearly, the relative density of composites was calculated according to the ratio of the measured bulk density to the theoretical density. The results are shown in Table 1.

Table 1. Density of the SiC_f/Al-Mg composite (g/cm^3).

Composite	SiC_f/Al	SiC_f/Al-5Mg	SiC_f/Al-10Mg	SiC_f/Al-15Mg	SiC_f/Al-20Mg
Bulk density	2.59	2.62	2.59	2.53	2.41
Relative density	96.64%	98.36%	98.27%	97.27%	94.21%

3.2. Phase Analysis

Figure 3 shows the XRD spectra of SiC_f/Al-Mg composites. It is clear from the spectra that each composite contains strong Al and SiC peaks, and an $Al_{12}Mg_{17}$ peak. The SiC peaks are situated at 2θ of 35.7°, 60.4°, and 71.8°, respectively. According to JCPDS cards (SiC: 49-1428), the three SiC peaks correspond to the (111), (220), and (311) crystallographic planes of β-SiC. The absence of the diffraction peaks of Mg in the XRD patterns might be attributed to the solid solution that formed with the Al matrix during the fabrication of the composites [23].

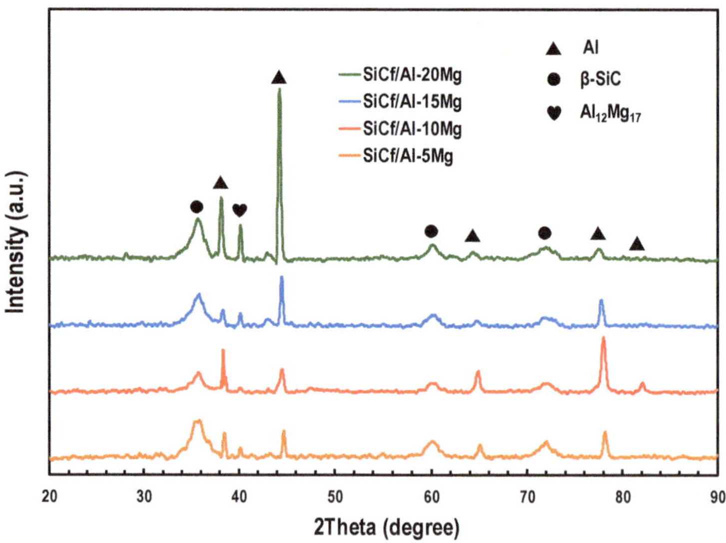

Figure 3. XRD patterns of the SiC_f/Al-Mg composites.

According to Bragg's law, solid solution atoms in Al which are smaller than Al atoms will shift the position of the diffraction peaks of Al towards the lower diffraction angle. From Figure 3, it could be seen that the position of the Al peaks shifted to the low diffraction angle when the Mg element was introduced. Therefore, it can be inferred that Mg was dissolved into the Al matrix during the fabrication process. The existence of $Al_{12}Mg_{17}$ peaks demonstrated that Mg reacted with Al, and it can be seen that the $Al_{12}Mg_{17}$ peak in the pattern of SiC_f/Al-20Mg was more prominent, which implies that more $Al_{12}Mg_{17}$ was generated in SiC_f/Al-20Mg. It has been reported that SiC may react with the Al matrix at a high temperature to produce the reactants Al_4C_3 and Si [24]. However, in the current work, no XRD patterns showed the presence of the Al_4C_3 and Si phases. As we know, the Al_4C_3 phase is a brittle phase, and its presence is not conducive to the mechanical properties of the composite. This is probably due to the low preparation temperature, short processing time, and the existence of Mg. All these factors would restrict the reaction of SiC fiber with Al. Similar results were also found in the literature [20].

In order to further analyze the distribution of different elements in the SiC_f/Al-Mg composites, the selected areas of SiC_f/Al-20Mg composites were examined by EDS surface scanning, as shown in Figure 4. Figure 4a shows the selected area and Figure 4b–d show the distribution of each element. It is clear from Figure 4b,c that the Mg co-existed with the Al element in the same area, implying that Mg was diffused into the Al matrix and formed a solid solution. This is consistent with the analysis of the XRD pattern (Figure 3). On the other hand, the distribution of the Si element (Figure 4d) matched with the profile of SiC fibers and no Si elements were present in the matrix. As a result, this may also indicate that the reaction between the SiC fiber and the Al matrix during the composites' fabrication is not obvious.

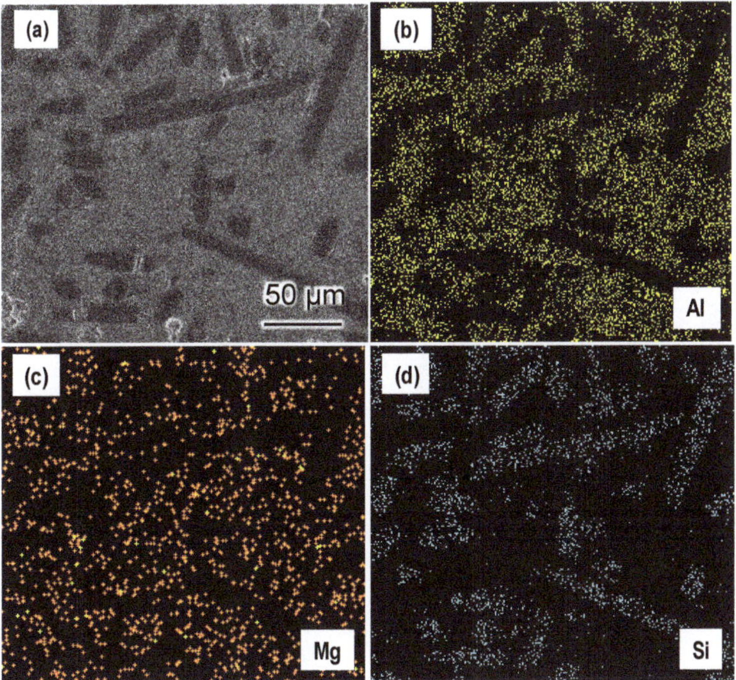

Figure 4. EDS analysis of the SiC_f/Al-20Mg composite: (**a**) SEM image of the selected area; (**b**) mapping of Al element; (**c**) mapping of Mg element; (**d**) mapping of Si element.

3.3. Mechanical Properties

3.3.1. Hardness

Figure 5 shows the Vickers hardness of SiC_f/Al-Mg composites. The Vickers hardness of the composites increased with the increase in the mass fraction of Mg.

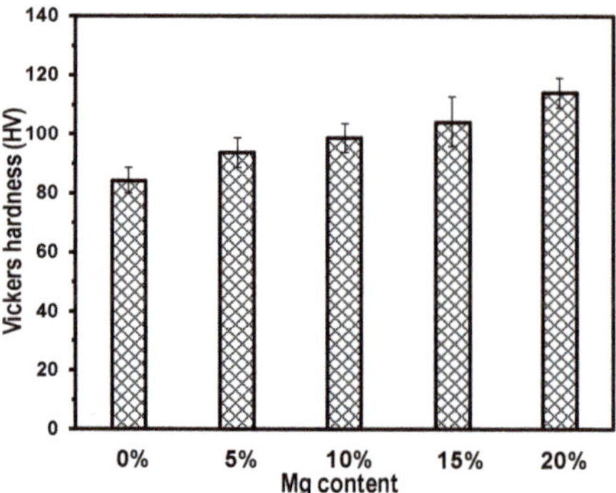

Figure 5. Vickers hardness of composites.

The Vickers hardness of the SiC$_f$/Al-20Mg composite was 114.06 HV, which was 35.56% higher than that of SiC$_f$/Al (84.14 HV). Obviously, the addition of Mg could improve the composite's hardness due to the solution-strengthening effect [25]. Moreover, the precipitate phase Al$_{12}$Mg$_{17}$ could also play a role of dispersion strength.

3.3.2. Flexural Strength

Figure 6 shows the stress–strain curves of different composites. The SiC$_f$/Al composite exhibited the best plasticity, and the fracture strain was larger than 1.5% (Figure 6a). However, for the SiC$_f$/Al-Mg composites, the fracture elongation decreased with the increase in Mg content. To clarify the effect of the SiC fiber on the mechanical properties of the composites, Figure 6b shows a comparison of the Al-10Mg alloy and the SiC$_f$/Al-10Mg composite. The SiC$_f$/Al-10Mg composites showed considerably higher flexural strength than those of the Al-10Mg alloy, while showing less apparent elongation. This might be attributed to the strong fiber–matrix interface bonding. SiC fibers are capable of carrying a large load, but they have a small deformation.

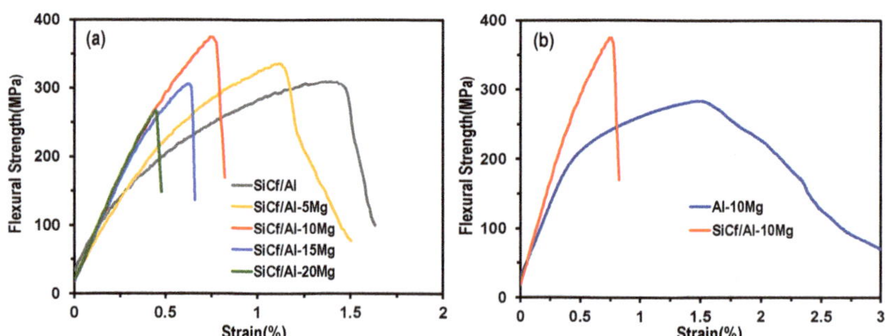

Figure 6. Mechanical properties of different composites: (**a**) stress–strain curves of SiC$_f$/Al-Mg composites; (**b**) comparison of stress–strain curves between Al-10Mg alloy and the SiC$_f$/Al-10Mg composite.

Table 2 summarizes the mechanical properties of SiC$_f$/Al-Mg composites. It was found that the modulus of the SiC$_f$/Al-Mg composites increased with the increase in Mg content, firstly. Moreover, it remained constant when the Mg content was beyond 10%. The modulus of the SiC$_f$/Al-20Mg composite was 162.6 GPa, which is 41.6% higher than that of SiC$_f$/Al. The interfacial bonding between the SiC fibers and the aluminum matrix can be strengthened by the addition of the Mg element, allowing for the full utilization of the

SiC fibers' load-bearing capability [19]. This meant that the composite's modulus could be improved by the addition of Mg with a suitable content.

Table 2. Mechanical properties of the SiC$_f$/Al-Mg composites.

Composite	Young's Modulus (GPa)	Flexural Strength (MPa)
SiC$_f$/Al	114.8 ± 3.6	309 ± 9
SiC$_f$/Al-5Mg	125.1 ± 1.0	324 ± 12
SiC$_f$/Al-10Mg	161.7 ± 2.0	372 ± 16
SiC$_f$/Al-15Mg	160.3 ± 1.7	331 ± 10
SiC$_f$/Al-20Mg	162.6 ± 5.6	283 ± 19

Compared to our previous work [22], the Young's modulus of the SiC$_f$/Al composites was much higher that of the Al matrix composites reinforced with SiC$_p$ and C$_f$ when the volume fraction of reinforcement was similar. Therefore, it can be considered that the introduction of Mg is conducive to improve the stiffness of Al matrix composites, with a better damping capacity expected. Again, it can be seen that the load-bearing capability of the SiC fiber was maximized when the Mg content exceeded 10%, and thus the modulus did not continue to increase with increasing Mg content. Correspondingly, the best flexural strength was achieved for the SiC$_f$/Al-10Mg composites with a value of 372 MPa. The SiC fibers were evenly distributed throughout the matrix. The Mg element not only strengthened the interface bonding, but also the aluminum matrix, thus improving the flexural strength and the elastic modulus of the composite. However, when the Mg content was high, the pores started to form, which made the composites more easily breakable when subjected to external forces.

3.4. Fracture Morphology Analysis

In order to better understand the effect of Mg content on the mechanical properties of SiC$_f$/Al-Mg composites, the fracture morphologies of SiC$_f$/Al-Mg composites were examined (Figure 7).

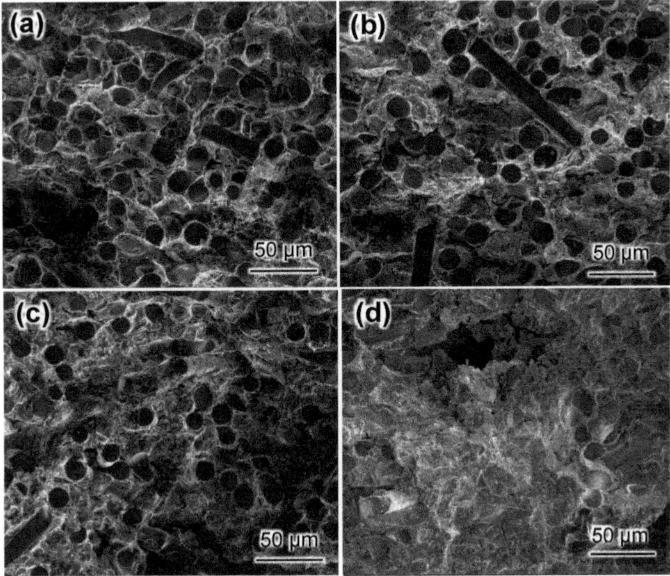

Figure 7. Fracture morphologies of the composites: (**a**) SiC$_f$/Al-5Mg; (**b**) SiC$_f$/Al-10Mg; (**c**) SiC$_f$/Al-15Mg; (**d**) SiC$_f$/Al-20Mg.

It could be seen that the fracture surfaces were uneven and the fibers were still embedded in the matrix, indicating that a strong interface formed between the fibers and the matrix. When the Mg content was low, some tearing edges, which are the symbol of ductile facture mode, were observed in SiC_f/Al-5Mg, as shown in Figure 7a. When the Mg mass fraction was 10%, as shown in Figure 7b, it was found that the composite's fracture surface presented few dimples, which suggests that the SiC fibers bonded strongly with the aluminum matrix. The strong interfacial bonding facilitated the transfer of loads. As a result, the composite exhibited excellent mechanical properties. As seen in Figure 7c,d, the composite fracture surfaces had porous flaws, which may reduce the mechanical capabilities of the composites. In particular, the pores were more obvious for the SiC_f/Al-20Mg composite. Such pores would act as defects and cause stress concentration when the load was applied. If the stress intensity was very high, the cracks would first be initiated from the defects. Meanwhile, due to a strong interfacial bond between the fibers and matrix, the cracks' propagation would not be inhibited and would penetrate directly into the fiber. As a result, low-stress damage occurred in the composites, as shown in Figure 6.

3.5. Damping Behavior

3.5.1. Damping Capacity at Room Temperature

The damping capacity as a function of strain is depicted in Figure 8. The damping capacity of all composites exhibited a weak dependence on the strain when it was lower than 0.001%. And then the damping capacity increased dramatically with an increase in the strain (higher than 0.001%). At a strain of 0.001%, the Q^{-1} values of SiC_f/Al-5Mg, SiC_f/Al-10Mg, and SiC_f/Al-15Mg were around 0.003, respectively, and were about 0.004 for the SiC_f/Al-20Mg. At a strain of 0.01%, all of the composites exhibited a Q^{-1} value of 0.007~0.011, which was higher than that for the alloys with similar composition [21], demonstrating that the SiC fiber had a positive effect on improving the damping capacity of the composite.

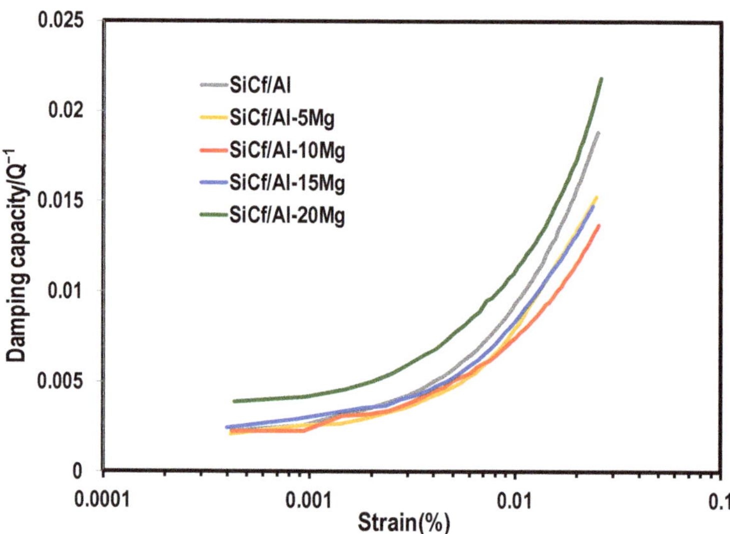

Figure 8. Strain dependence of the damping capacity for the SiC_f/Al-Mg composites at room temperature.

For the metal matrix composites at a low temperature regime, the dominant damping mechanisms could be associated with the consisting phases and the dislocation in the composite [26,27]. The contribution of the consisting phases to the damping can be described by the rule of mixing (ROM). The damping values of Al and SiC were 0.003 and 0.001 [19,28], respectively. The damping values of the SiC_f/Al composite calculated by ROM were less than 0.003 in most of the test ranges of strain, which was lower than the experimental ones. This implies that the damping of the SiC_f/Al composite should be related to other mechanisms. The large difference in CTE between the SiC fiber and the Al matrix would cause the accumulation of thermal stresses during the fabrication process, which would lead to the creation of dislocations in the matrix. It was considered that the dislocation also played a role in the damping capacity of SiC_f/Al composites at RT. As for dislocation damping, the Granato–Lücke (G-L) model is widely accepted [29].

According to the G-L model, the dislocation in the material was similar to the elastic strings pinned in the weak points (such as solution atoms, vacancies, etc.) and the strong points (such as network nodes of dislocation, grain boundaries, the secondary phases, etc.). The dislocations sweep these pinning points and dissipate the vibration energy into the thermal energy irreversibly, which is the reason for the occurrence of dislocation damping. At a low strain amplitude, most of the dislocations were pinned by the weak pinning points and only oscillated in a small area, and thus the damping capacity was relatively low. When the strain reached a certain level, the snow-like breakaway of dislocations from the weak pinning point would happen. Thus, the area swept by dislocation segments would become larger. Consequently, when the strain was over a critical value, the damping capacity of the material increased dramatically with the increase in strain amplitude.

According to the G-L model, at a low strain amplitude, damping is dependent on the frequency and the strain. The strain-independent damping component symbolized as Q_0 can be described by Equation (1), as follows [21]:

$$Q_0 \sim \rho L_c{}^4 \tag{1}$$

where ρ is the mobile dislocation density and L_c represents the average length of the dislocation segment between weak pinning points.

Mg atoms, as solute atoms, affected the damping capacity of the SiC_f/Al-Mg composites in two main aspects. Firstly, Mg atoms acted as weak pinning points, which hindered the bowing movement of the dislocations. Secondly, an increase in Mg content led to a higher dislocation density in the matrix. According to Formula (1), when the strain amplitude is low, the strain-independent damping of the composites depends on the dislocation density in the matrix and the average distance between weak pinning points. As the Mg content in the matrix increases, a large number of dislocations are generated in the matrix, resulting in dislocation density ρ increasing. Meanwhile, the increase in solute atoms in the matrix will also shorten the average distance L_c of weak pinning points. The strain-independent damping capacity of the composites will be determined by the combined effect of these two factors.

From Figure 8, it can be seen that the damping capacity at a low strain amplitude for the composites with Mg content of 0%, 5%, 10%, and 15% was roughly similar. This proves that the damping capacity is not significantly affected by the increase in Mg content through the above-mentioned combined effect. However, it is worth noting that the strain-independent damping capacity of the SiC_f/Al-20Mg composite was significantly improved compared to other composites. It was suggested by previous research [30] that the presence and homogeneous distribution of a large number of secondary phases were beneficial to improve the damping capacity of aluminum alloys at RT, due to the generation of a large number of movable dislocations around it. In the SiC_f/Al-20Mg composite, abundant $Al_{12}Mg_{17}$ phases precipitated in the matrix, and were detected by XRD. It is reasonably believed that the $Al_{12}Mg_{17}$ phase may make a contribution to the improvement of strain-independent damping in the SiC_f/Al-20Mg composite.

On the other hand, the strain dependent damping component Q_H conforms to formulae as follows [19]:

$$Q_H^{-1} = \frac{C_1}{\varepsilon} \exp\left(-\frac{C_2}{\varepsilon}\right) \tag{2}$$

$$C_1 = \frac{\rho F_B L_N^3}{6bEL_C^2} \tag{3}$$

$$C_2 = \frac{F_B}{bEL_C} \tag{4}$$

where ε represents the strain amplitude; C_1 and C_2 are constants related to the properties of a material; ρ signifies the dislocation density; F_B is the binding force between dislocations and weak pinning points; E is the elastic modulus; L_C and L_N are the average dislocation distance between the weak pinning points and the strong pinning points, respectively; and b is the Burger's vector. Through Equation (2), the formula can be transformed into the following form:

$$\ln\left(\varepsilon Q_H^{-1}\right) = \ln C_1 - \frac{C_2}{\varepsilon} \tag{5}$$

That is, if the strain-dependent damping of composites follows the G-L model, the plot of $\ln\left(\varepsilon Q_H^{-1}\right)$ versus $\frac{1}{\varepsilon}$ should be satisfied with a linear relationship. In this work, the $\ln\left(\varepsilon Q_H^{-1}\right) - \frac{1}{\varepsilon}$ plots obtained from the experiments are presented in Figure 9. It is clear that the favorable linear fit is exhibited. Hence, G-L theory is conformed for the composites in this work.

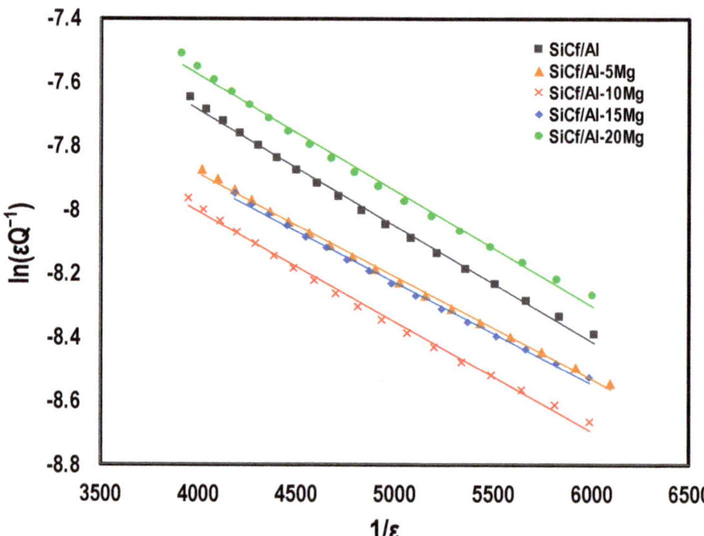

Figure 9. G-L plots of strain-dependent damping in SiC_f/Al-Mg composites.

As the strain amplitude increased, the dislocations gradually began to break away from the weak pinning points and only engaged in a bowing motion within strong pinning points. From Formula (2), it can be seen that the strain-dependent damping increased with the average distance between the strong pinning points. For SiC_f/Al-Mg composites, when the SiC_f content was constant, the distance between the strong pinning points was mainly dependent on the grain size of the matrix. However, an increase in Mg content would reduce the grain size of the matrix [31], thereby reducing the strain-dependent damping capacity of the composite. In the current work, for SiC_f/Al-5Mg, SiC_f/Al-10Mg, and SiC_f/Al-15Mg composites, the dominant reason for the damping reduction could be attributed to the grain size refinement, as shown in Figure 8. However, the SiC_f/Al-20Mg still exhibited a higher damping capacity than that of the SiC_f/Al composite. This might be attributed to the effect of the second phase, mentioned earlier on, enhancing the damping mechanisms. In addition, a certain amount of flaws contained in SiC_f/Al-20Mg composites may also contribute to the enhanced damping capacity.

3.5.2. Damping Capacity at Elevated Temperatures

Figure 10 shows the temperature-dependent damping capacity of the SiC_f/Al-Mg composites. Obviously, the damping capacity for all composites exhibited a tendency for monotonic increase with the testing temperature. Such a phenomenon was also observed in other works [32–34]. Furthermore, the damping capacity of SiC_f/Al-Mg composites is lower than that of SiC_f/Al and shows no significant increase until 200 °C. This may be attributed to the second phase, which formed more strong pinning points and increased the dislocation energy needed to break away from new pinning points.

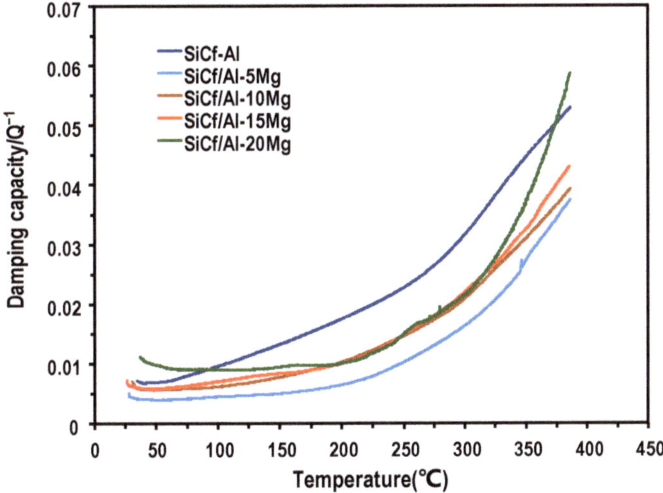

Figure 10. Temperature dependence of damping capacity in SiC_f/Al-Mg composites.

Furthermore, there was no evident damping peak in the curve of composites with low Mg content. Notably, SiC_f/Al-15Mg exhibited one damping peak at 130 °C and SiC_f/Al-20Mg exhibited two damping peaks at 165 °C and 262 °C, respectively. According to a previous work [31], the damping peak in metal matrix composites appearing at low-temperature regimes was associated with dislocation damping, while it was explained by grain boundary damping and interface damping mechanisms at high-temperature regimes (>200 °C). It was found that Mg introduction could decrease the stacking fault energy markedly and create dislocations and stacking faults in the Al matrix [21]. Thus, the occurrence of the damping peaks in SiC_f/Al-15Mg and SiC_f/Al-20Mg composites could be associated with the increase in the dislocation density caused by the high Mg content. The dislocation was the source of the damping capacity.

Moreover, the introduction of Mg refined the grain size of the matrix and led to an increase in the grain boundary area. When the temperature rose up to a high level, a softening of the matrix happened and viscous sliding between grain boundaries occurred. The grain boundary sliding dissipated more friction energy into thermal energy and led to a pronounced increase in the damping capacity, as well as the formation of a damping peak. It was also found in Figure 10 that the damping capacity increased with increasing Mg content when the temperatures were below 200 °C and above 300 °C. Based on the above analyses, the increase in the damping capacity of composites with Mg content with a low-temperature regime might be associated with the increased dislocations and stacking faults; the increase in the damping capacity of composites with Mg content at high-temperature regimes might be driven by the increased grain boundary area.

4. Conclusions

SiC-fiber-reinforced Al-Mg matrix composites with various levels of Mg content are fabricated via combining colloidal dispersion with a squeeze melt infiltration process. The effect of Mg content on the mechanical properties and damping capacity are investigated. The results are summarized as follows:

(1) SiC_f/Al-Mg composites exhibit a homogeneous distribution of SiC fibers and a high relative density (higher than 97%) when the mass fraction of Mg is less than 20%. Mg could dissolve into the Al matrix, forming the $Al_{12}Mg_{17}$ precipitate phase. Fibers are well bonded with the Al-Mg matrix and no obvious reactive phase is present at the fiber–matrix interface.

(2) The Vickers hardness of the composites increases with increasing Mg content, and the highest value is 114.06 HV for SiC_f/Al-20Mg, which is 35.56% higher than that of SiC_f/Al. The enhanced hardness relates to the strengthening effect caused by the introduction of Mg.

(3) SiC_f/Al-10Mg has the best flexural strength and elastic modulus, 372 MPa and 161.7 GPa, but the fracture elongation of the composites decreases with the increase in Mg content. This could be attributed to the strengthened interfacial bonding by the introduction of the Mg element.

(4) The damping capacity of SiC_f/Al-Mg shows a weak dependence on the strain when the strain amplitude is lower than 0.001%. And then the damping capacity increases dramatically with an increase in the strain (higher than 0.001%), which is better than the alloys with similar composition, demonstrating that the SiC fiber has a positive effect on improving the damping capacity of the composite.

(5) The temperature-dependent damping capacity of the SiC_f/Al-Mg composites reveals that all composites exhibited a tendency towards monotonic increase with testing temperature. Such an increase is more obvious at temperatures beyond 200 °C. This is attributed to the second phase, which forms more strong pinning points and increases the dislocation energy needed to break away from new pinning points.

Author Contributions: Conceptualization, J.S.; methodology, J.S. and G.L.; software, G.L.; validation, J.S.; formal analysis, Y.Z.; investigation, G.L. and C.S.; resources, J.S.; data curation, G.L.; writing—original draft preparation, G.L.; writing—review and editing, J.S., Z.L., Y.Z. and C.S.; visualization, G.L.; supervision, J.S.; project administration, J.S.; funding acquisition, J.S. and J.D. All authors have read and agreed to the published version of the manuscript.

Funding: This work was supported by the Joint Funds of the National Natural Science Foundation of China and the China Academy of Engineering Physics (NSFA, Grant No.: U1630129).

Institutional Review Board Statement: Not applicable.

Informed Consent Statement: Not applicable.

Data Availability Statement: The data presented in this study are available on request from the corresponding author due to privacy.

Conflicts of Interest: The authors declare that the research was conducted in the absence of any commercial or financial relationships that could be construed as potential conflicts of interest.

References

1. Vineet, C.; Himadri, C.; Dora, T.L. A review on fabrication methods, reinforcements and mechanical properties of aluminum matrix composites. *J. Manuf. Processes.* **2020**, *56*, 1059–1074. [CrossRef]
2. Alam, M.; Ya, H.; Azeem, M.; Mazli, M.; Mohammad, Y.; Faisal, M.; Roshan, V.M.; Salit, M.S.; Akhter, H.A. Advancements in aluminum matrix composites reinforced with carbides and graphene: A comprehensive review. *Nanotech. Rev.* **2020**, *56*, 1059–1074. [CrossRef]
3. Choy, K.L. Effects of surface modification on the interfacial chemical stability and strength of continuous SiC fibers after exposure to molten aluminium. *Scr. Metall. Mater.* **1995**, *32*, 219–224. [CrossRef]
4. Ha, R.; Xiao, W.K. Development Status of Short Fiber Reinforced Aluminum Matrix Composite. *Hot Working Technol.* **2022**, *51*, 10–13. [CrossRef]

5. Ishikawa, T. Polymeric and inorganic fibers. *Materials* **2009**, *246*, 1097–1104.
6. Bunsell, A.R.; Piant, A. A review of the development of three generations of small diameter silicon carbide fibres. *Mater. Sci.* **2006**, *41*, 823–839. [CrossRef]
7. Shirvanimoghaddam, K.; Hamim, S.U.; Karbalaei, A.M.; Seyed, M.F.; Hamid, K.; Amir, H.P.; Ehsan, G.; Mahla, Z.; Khurram, S.M.; Shian, J.; et al. Carbon fiber reinforced metal matrix composites: Fabrication processes and properties. *Compos. Part A Appl. Sci. Manuf.* **2017**, *92*, 70. [CrossRef]
8. Rams, J.; Urena, A.; Escalera, M.D.; Sanchez, M. Electroless nickel coated short carbon fibres in aluminium matrix composites. *Compos. Part A* **2007**, *38*, 566–575. [CrossRef]
9. Singha, B.; Balasubramanian, M. Processing and properties of copper-coated carbon fibre-reinforced aluminium alloy composites. *J. Mater. Process. Technol.* **2009**, *209*, 2104–2110. [CrossRef]
10. Updike, C.A.; Bhagat, R.B.; Pechersky, M.J.; Amateau, M.F. The damping performance of aluminum based composites. *JOM* **1990**, *42*, 42–46. [CrossRef]
11. Morelli, E.C.; Urreta, S.E.; Schaller, R. Mechanical spectroscopy of thermal stress relaxation at metal-ceramic interfaces in aluminium-based composites. *Acta Mater.* **2000**, *48*, 4725–4733. [CrossRef]
12. Singh, S.; Pal, K. Effect of texture evolution on mechanical and damping properties of $SiC/ZnAl2O4/Al$ composite through friction stir processing. *J. Market. Res.* **2019**, *8*, 222–232. [CrossRef]
13. Theo, O.J.; Kenneth, K.A.; Michael, O.B. On the microstructure, mechanical behaviour and damping characteristics of Al-Zn based composites reinforced with martensitic stainless steel (410L) and silicon carbide particulates. *Int. J. Lightweight Mater. Manuf.* **2022**, *5*, 279–288. [CrossRef]
14. Bhagat, R.B.; Amateau, M.F.; Smith, E.C. Damping Behavior of Planar Random Carbon Fiber Reinforced 6061 Al Matrix Composites Fabricated by High-Pressure Infiltration Casting. *J. Compos. Tech. Res.* **1989**, *11*, 113–116. [CrossRef]
15. Stanley, E.N.; Vinod, B.; Ramakrishna, A. Energy Dissipation Behaviour of Bamboo Leaf Ash reinforced Aluminium Metal Matrix Composites. *J. Inst. Eng. India Ser.* **2022**, *103*, 149–155. [CrossRef]
16. Reddy, K.V.; Naik, R.B.; Rao, G.R.; Reddy, G.M.; Kumar, R.A. Microstructure and damping capacity of AA6061/graphite surface composites produced through friction stir processing. *Comp. Commun.* **2020**, *20*, 100352. [CrossRef]
17. Alaneme, K.K.; Jafaar, L.; Bodunrin, M.O. Structural characteristics, mechanical behaviour and Damping properties of Ni modified high zinc Al-Zn alloys. *Mater. Res. Express.* **2019**, *6*, 1065–1671. [CrossRef]
18. Xu, Y.J. *Effect of Matrix Alloy on Microstructure and Mechanical Properties of Continuous SiC_f/Al Composites*; Nan-chang Hangkong University: Nanchang, China, 2014; p. 42.
19. Chu, D.S.; Ma, Y.; Li, P.; Huang, H.; Tang, P.J. Microstructure and Damping Behavior of Continuous W-Core-SiC Fiber-Reinforced Aluminum Matrix Composites. *Appl. Compos. Mater.* **2021**, *28*, 1631–1651. [CrossRef]
20. Geng, L.; Zhang, H.W.; Li, H.Z. Effects of Mg content on microstructure and mechanical properties of $SiC_p/Al-Mg$ composites fabricated by semi-solid stirring technique. *Trans. Nonferrous Met. Soc. Chin.* **2010**, *20*, 1851–1855. [CrossRef]
21. Li, Z.; Li, X.; Yan, H.; Guan, L.N.; Huang, L.J. Achieving high damping and excellent ductility of Al-Mg alloy sheet by the coupling effect of Mg content and fine grain structure. *Mater. Charact.* **2021**, *174*, 110974. [CrossRef]
22. Lv, Z.Z.; Sha, J.J.; Lin, G.Z.; Wang, J.; Guo, Y.C.; Dong, S.Q. Mechanical and thermal expansion behavior of hybrid aluminum matrix composites reinforced with SiC particles and short carbon fibers. *J. Alloys Compd.* **2023**, *947*, 169550. [CrossRef]
23. Liu, Z.; Sun, J.; Yan, Z.; Lin, Y.J.; Liu, M.P.; Hans, J.R.; Arne, K.D. Enhanced ductility and strength in a cast Al-Mg alloy with high Mg content. *Mater. Sci. Eng. A Struct. Mater. Prop. Microstruct. Process.* **2021**, *140*, 806. [CrossRef]
24. Mitar, R.; Chalapathi, R.V.S.; Maiti, R.; Chakraborty, M. Stability and response to rolling of the interfaces in cast $Al-SiC_p$ and $Al-Mg$ alloy-SiC_p composites. *Mater. Sci. Eng. A* **2004**, *379*, 391–400. [CrossRef]
25. Jeong, H.T.; Kim, W.J. Strain hardening behavior and strengthening mechanism in Mg-rich Al–Mg binary alloys subjected to aging treatment. *Mater. Sci. Eng. A* **2020**, *794*, 139862. [CrossRef]
26. Zhang, Y.; Ma, N.; Li, X.; Wang, H. Study on damping capacity aluminum composite reinforced with in situ TiAl3 rod. *Mater. Des.* **2008**, *29*, 1057–1066. [CrossRef]
27. Zhang, Y.; Ma, N.; Wang, H.; Le, Y.; Li, X. Damping capacity of in situ TiB2 particulates reinforced aluminium composites with Ti addition. *Mater. Des.* **2007**, *28*, 628–650. [CrossRef]
28. Madeira, S.; Carvalho, O.; Carneiro, V.H.; Soares, D.; Silva, F.S.; Miranda, G. Damping capacity and dynamic modulus of hot pressed AlSi composites reinforced with different SiC particle sized. *Compos. Part B Eng.* **2016**, *90*, 399–405. [CrossRef]
29. Granato, A.V.; Lucke, K. Theory of Mechanical Damping Due to Dislocations. *J. Appl. Phys.* **1956**, *27*, 583–593. [CrossRef]
30. Jiang, H.J.; Liu, C.Y.; Zhang, B.P.; Xue, P.; Ma, Z.Y.; Luo, K.; Ma, M.Z.; Liu, R.P. Simultaneously improving mechanical properties and damping capacity of Al-Mg-Si alloy through friction stir processing. *Mater. Character.* **2017**, *131*, 425–430. [CrossRef]
31. Li, Z.; Yan, H.; Chen, J.; Xia, W.J.; Zhu, H.M.; Su, B.; Lia, X.Y.; Song, M. Enhancing damping capacity and mechanical properties of Al-Mg alloy by high strain rate hot rolling and subsequent cold rolling. *J. Alloys Compd.* **2022**, *908*, 164677. [CrossRef]
32. Subhash, S.; Kaushik, P. Influence of surface morphology and UFG on damping and mechanical properties of composite reinforced with spinel $MgAl_2O_4$-SiC core-shell microcomposites. *Mater. Char.* **2017**, *123*, 244–255. [CrossRef]

33. Subhash, S.; Keerti, R.; Kaushik, P. Synthesis, characterization of graphene oxide wrapped silicon carbide for excellent mechanical and damping performance for aerospace application. *J. Alloys Compd.* **2018**, *740*, 436–445. [CrossRef]
34. Girish, B.M.; Prakash, K.R.; Satish, B.M.; Jain, P.K.; Prabhakar, P. An investigation into the effects of graphite particles on the damping behavior of ZA-27 alloy composite material. *Mater. Des.* **2011**, *32*, 1050–1056. [CrossRef]

 *materials*

MDPI

Article

Improved Bending Strength and Thermal Conductivity of Diamond/Al Composites with Ti Coating Fabricated by Liquid–Solid Separation Method

Hongyu Zhou [1,*], Qijin Jia [2], Jing Sun [3], Yaqiang Li [4], Yinsheng He [1], Wensi Bi [5] and Wenyue Zheng [1,*]

[1] National Center for Materials Service Safety, University of Science and Technology Beijing, Beijing 100083, China; heyinsheng@ustb.edu.cn
[2] Beijing System Design Institute of Electro-Mechanic Engineering, Beijing 100039, China; jqj627@sina.com
[3] Beijing Hangxing Machinery Co., Ltd., Beijing 100013, China; winlxc2@sina.com
[4] Institute for Advanced Materials and Technology, University of Science and Technology Beijing, Beijing 100083, China; yqli0677@163.com
[5] National Academy of Forestry and Grassland Administration, Beijing 102600, China; biwensi2008@163.com
* Correspondence: hyzhou@ustb.edu.cn (H.Z.); zheng_wenyue@ustb.edu.cn (W.Z.);
  Tel.: +86-010-62334839 (H.Z.)

Abstract: In response to the rapid development of high-performance electronic devices, diamond/Al composites with high thermal conductivity (TC) have been considered as the latest generation of thermal management materials. This study involved the fabrication of diamond/Al composites reinforced with Ti-coated diamond particles using a liquid–solid separation (LSS) method. The interfacial characteristics of composites both without and with Ti coatings were evaluated using SEM, XRD, and EMPA. The results show that the LSS technology can fabricate diamond/Al composites without Al_4C_3, hence guaranteeing excellent mechanical and thermophysical properties. The higher TC of the diamond/Al composite with a Ti coating was attributed to the favorable metallurgical bonding interface compounds. Due to the non-wettability between diamond and Al, the TC of un-coated diamond particle-reinforced composites was only 149 W/m·K. The TC of Ti-coated composites increased by 85.9% to 277 W/m·K. A simultaneous comparison and analysis were performed on the features of composites reinforced by Ti and Cr coatings. The results suggest that the application of the Ti coating increases the bending strength of the composite, while the Cr coating enhances the TC of the composite. We calculate the theoretical TC of the diamond/Al composite by using the differential effective medium (DEM) and Maxwell prediction model and analyze the effect of Ti coating on the TC of the composite.

Keywords: diamond/Al composite; liquid–solid separation (LSS); Ti coating; interfacial bonding; bending strength; thermal conductivity

Citation: Zhou, H.; Jia, Q.; Sun, J.; Li, Y.; He, Y.; Bi, W.; Zheng, W. Improved Bending Strength and Thermal Conductivity of Diamond/Al Composites with Ti Coating Fabricated by Liquid–Solid Separation Method. *Materials* **2024**, *17*, 1485.
https://doi.org/10.3390/ma17071485

Academic Editor: Alexander N. Obraztsov

Received: 20 February 2024
Revised: 18 March 2024
Accepted: 19 March 2024
Published: 25 March 2024

1. Introduction

The advent of advanced electronic devices such as insulated gate bipolar transistors (IGBTs), phased array radars, and high-power solid-state lasers has escalated the heat flux per unit area generated by chip computing, necessitating efficient heat dissipation for stable operation [1–3]. Higher TC than existing thermal management materials (such as Invar, Cu/W, Si/Al, SiC/Al, etc.) is needed to meet the urgent heat dissipation requirements of large-scale integrated circuits [4,5]. Although the TC of pure Al can reach 237 W/m·K, this is sufficient for conventional electronic packaging environments. But the coefficient of thermal expansion (CTE) of Al is 23.0×10^{-6}/K, which is too large to match that of the chip (i.e., $4.0{\sim}7.0 \times 10^{-6}$/K) [6]. This mismatch causes stress between the packaging material and the chip when the temperature changes, causing damage to the computing components [7]. The diamond with the CTE of $1.0{\sim}3.0 \times 10^{-6}$/K [8] composited to the

Al matrix to perfectly match the CTE of the chip. Diamond/Al composites emerge as promising candidates for advanced thermal management due to their matching CTE, superior TC, and low density [9,10].

However, the significant disparity in the physical and chemical characteristics between diamond particles and the Al matrix poses a challenge, leading to interface incompatibility that significantly reduces the TC of diamond/Al composite [11,12]. Furthermore, in conventional preparation methods (such as powder metallurgy [13], gas pressure infiltration [9], vacuum pressure infiltration [14], vacuum hot pressing [15], and spark plasma sintering [16]), the interface of the two phases inevitably produces Al_4C_3 under high-temperature or long-term contact conditions [9,13–16]. The presence of Al_4C_3 at the interface not only significantly decreases the TC of the composite [17] but also restricts its application due to the hydrolytic nature of Al_4C_3 [9,18]. Previously, it was believed that the diamond surface coating could enhance the interface bonding between the diamond and Al, improve TC, and avoid the formation of Al_4C_3. However, it has recently been found that even with an intact diamond coating, minor variations in preparation parameters could lead to the formation of Al_4C_3 [18]. Therefore, in order to ensure the stability of thermal management materials, the preparation method and process parameters must be strictly optimized and controlled.

An LSS technology has been developed based on the principles of powder metallurgy and semi-solid thixoforming [19,20], which has garnered significant attention. Characterized by a low heating temperature and brief holding time, LSS technology prevents the formation of Al_4C_3 with deteriorating properties at the interface [21]. Additionally, by leveraging the differential flowability of solid and liquid phases under stress, LSS enables the fabrication of diamond/Al composite shells with a graded distribution of diamond particles and thermophysical properties [22]. This graded distribution addresses the specific needs of different parts of the shell, enhancing the thermal stability of the integrated circuit module.

The TC of diamond/Al composites is primarily influenced by the interface bonding between the Al matrix and diamond particles. Impurities, holes, and other defects at the interface can induce phonon-boundary scattering, reducing the mean free path of phonons and then deteriorating the TC of the composite [4]. Diamond coating represents a cost-effective and efficient modification method, acting as a bridging agent between diamond and Al [23]. Our previous studies have explored the use of various coating layers, such as Cu [19], Ni [20], and Cr [21], to enhance interfacial bonding in LSS-fabricated diamond/Al composites. Recent studies show that Ti coating, a carbide-forming element, is adapted to form an interfacial layer between the matrix and reinforces and strengthens the interfacial bonding [24,25]. However, the influences of Ti coating on the mechanical properties and TC of diamond/Al composite fabricated by the LSS process have not been explored yet. In addition, there are no detailed studies to analyze the effects of composites fabricated with different coatings on bending strength.

This study employed vacuum ion plating to deposit a Ti coating on the surface of diamond particles, utilizing LSS to fabricate diamond/Al composites with a reinforcement phase volume of 40%. The interfacial characterization, corresponding bending property, and TC of the diamond/Al composites have been improved by the Ti coating. The theoretical TC of the diamond/Al composite was calculated using the DEM and Maxwell prediction model.

2. Materials and Methods

2.1. Materials

For this study, industrial-grade Al powder with a 99.81% purity (by mass) and an average particle size of 37 μm (Zhengzhou Aerospace Aluminum Co., Ltd., Zhengzhou, China) was used. Additionally, MBD-4 grade synthetic single-crystal diamond particles with an average size of 106 μm (Henan Huanghe Whirlwind Co., Ltd., Changge, China) were selected as the primary raw materials. The preferred alternative for diamond/Al composite interface layers is a nanoscale coating layer with high sound velocities, such as Ti, Cr, and other metals [26]. With the increase in coating thickness, the tensile, compressive, and bending strength of diamond/Al composites gradually increase [4], but the TC of the composite increases first and then decreases [27]. On the premise of improving the interface bonding, the coating thickness should be as low as possible to reduce the interface thermal resistance [24]. However, it is difficult to control the layer thickness below 50 nm. The investigation involved the purchase of Ti-coated diamond particles that were manufactured utilizing the vacuum ion plating process. Based on the expected duration of the coating process, the thickness of the Ti coating was approximately 100 nm.

2.2. Fabrication of Diamond/Al Composites

The volumetric fraction of diamond in this composite was 40%. Figure 1 depicts a schematic diagram illustrating the LSS process. The experimental procedure can be articulated as follows: Initially, the diamond particles and Al powder were mixed in a 1:4 volume ratio for 8 h using a Turbula Shaker/Mixer (Model T2C, Glen Mills, PA, USA) to ensure uniform distribution. Furthermore, the homogenously mixed powder was placed into a mold and subjected to compression to form a blank using a cold-pressing technique in a four-column press (Model YQ28-100, Wodda, Zaozhuang, China) at a pressure of 300 MPa for 1 min. The blank measured 6.6 × 38 × 48 mm. Subsequently, the blank was moved to a custom-made LSS mold system and heated until it reached a stage where the liquid and solid components were blended and melted together. The detailed heating process was as follows: the mixture was initially heated at a rate of 20 °C/min to 450 °C and held for 20 min, and then heated for a second time at the same rate to 683 °C, where it was maintained for 40 min. In the fourth step, the molten metal was squeezed into the liquid chamber via the 2 mm LSS channel using a piston under a pressure of 60 MPa, and the diamond particles were completely trapped in the LSS chamber. Ultimately, under the action of cooling water, the slurry in the LSS chamber solidified layer by layer to form a diamond/Al composite with a dimension of 3 × 40 × 50 mm. Throughout the LSS process, hydrogen gas was employed to prevent the oxidation of the Al powder.

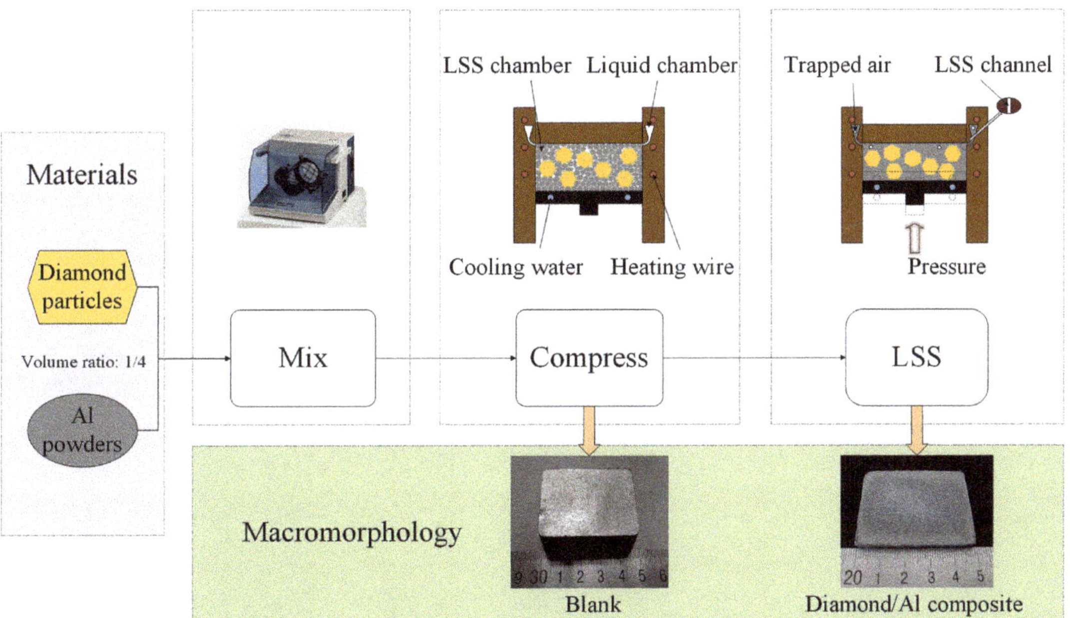

Figure 1. Schematic diagram of the LSS process.

2.3. Characterization

The diamond/Al composites underwent processing using a laser cutting machine and a diamond wheel grinder. The three-point bending strength of the composites with a dimension of 3 × 4 × 25 mm was examined by an RGM-3010 electronic universal testing machine (Shenzhen, China) at room temperature. The morphologies of the diamond particles, the surface structure, and the three-point bending fractography of the diamond/Al composites were observed by an EVO-18 scanning electron microscope (SEM, Zeiss, Oberkochen, Germany). The distributions of elements across the coating were determined by a JXA-8230 electron microprobe analyzer (EMPA, JEOL, Tokyo, Japan). The phase composition of the composites was certified by an Advance D8 X-ray diffractometer (XRD, Bruker, Saarbruecken, Germany) using Cu Ka radiation at 40 kV and 35 mA. The 2θ scans were executed between 20° and 80° at a scanning speed of 4°/min. In addition, the presence of Al_4C_3 was further confirmed by careful scanning between 30° and 45° at a scanning speed of 0.25°/min. Impurity elements such as N, H, and B are often present in diamond particles, with N being the most common. The N content of diamond particles was evaluated by a Nicolet iN10 MX Fourier transform infrared spectrometer (RTIR, ThermoFisher, Waltham, MA, USA). The thermal diffusion (α) of the composites with dimensions of φ 12.7 × 3 mm was measured by an LFA laser flash thermophysical machine (Netzsch, Selb, Germany) at room temperature. The TC value of composites was obtained by an equation: TC = ρ × α × C_p, where ρ (density) was determined by the Archimedes principle and C_p (specific heat capacity) was obtained based on the theoretical calculation. The C_p of the composite is equal to the respective C_p of the matrix and reinforcement phase multiplied by the corresponding volume fractions.

3. Results and Discussion

3.1. Microstructure of Diamond/Al Composites

The morphologies of the diamond particles without or with Ti coating are shown in Figure 2. Figure 2a,b illustrates the presence of partially fragmented particles in both types of diamond particles, and Figure 2c,d demonstrates the surface defects (shown by blue arrows) of the diamond particles that have been entirely covered by Ti coating. The metal coating could diffuse into the Al matrix and then form intermetallic compounds with Al, confirmed in diamond/Al with a TiC coating [26]. The Ti coating can improve the interfacial bonding force and enhance the interface conductivity, which is an effective way to fabricate diamond/matrix composites with high TC.

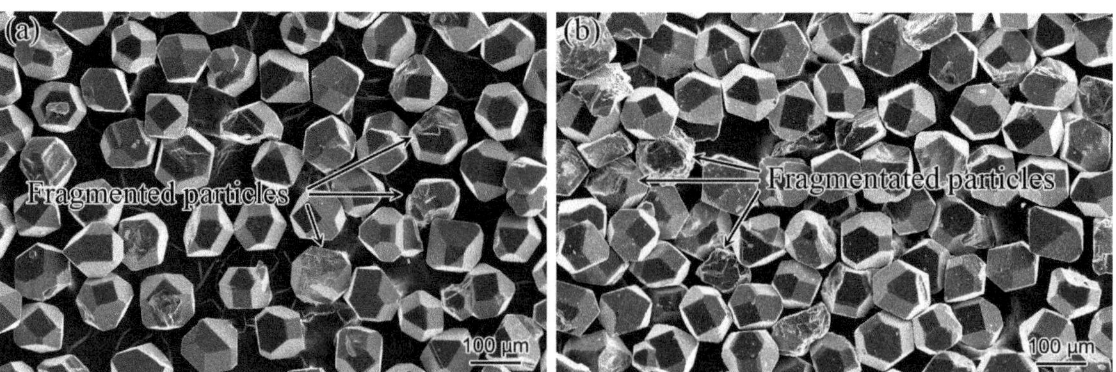

Figure 2. *Cont.*

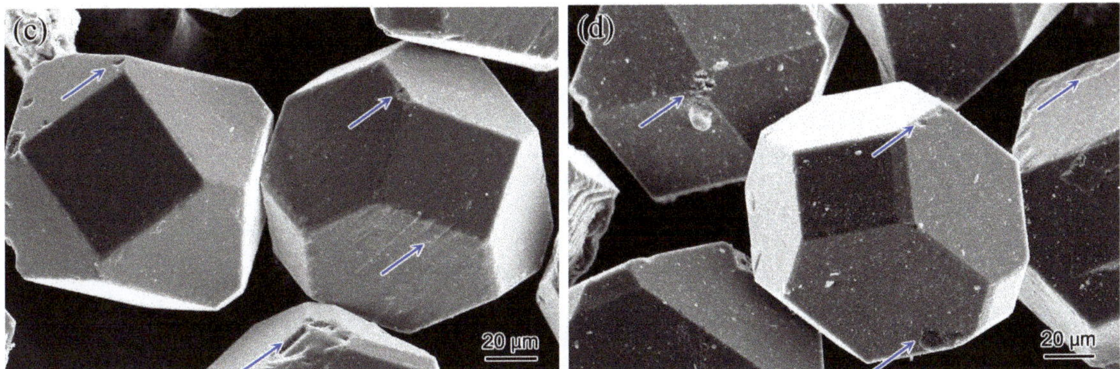

Figure 2. Morphologies of the diamond particles: (**a**) without Ti coating; (**b**) with Ti coating; (**c**) with surface defects; (**d**) with surface defects covered by Ti coating (the defects of diamond surface as shown by blue arrows).

3.2. Microstructure of Diamond/Al Composites

The microstructure of the diamond/Al composite fabricated by LSS technology is presented in Figure 3a. The synthetic diamond particles without Ti coating were uniformly distributed in the Al matrix, and no spalling of diamond particles was viewed. However, the microstructure of the composite cannot elucidate the behavior of interfacial bonding. Figure 3b displays the microstructure of the separated liquid phase. It can be seen that no diamond particles were discovered in the separated liquid phase, indicating the LSS channel can eliminate the spillover of diamond particles. This also verifies that the LSS technology can precisely fabricate diamond/Al composites with varying volume fractions of reinforcement phase by adjusting the size of the LSS chamber and liquid chamber (as depicted in Figure 1).

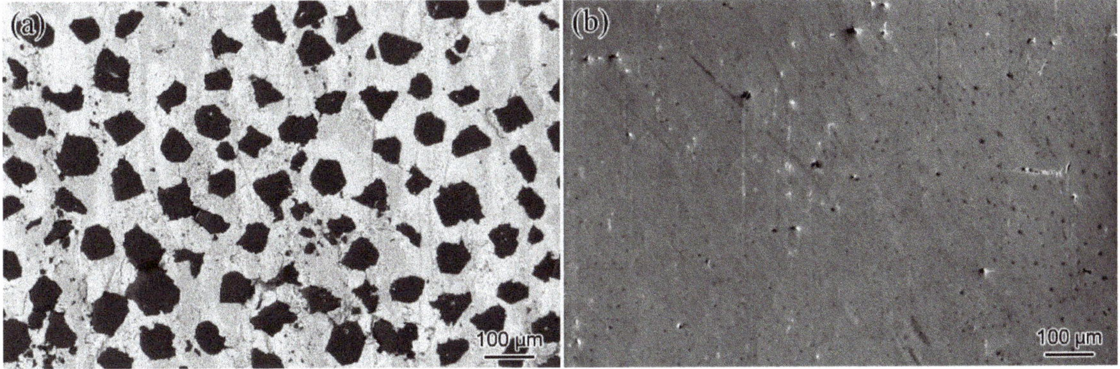

Figure 3. Microstructure of materials fabricated by LSS process: (**a**) diamond/Al composite; (**b**) separated liquid phase.

Fractography can determine whether the bonding state at the interface is of a mechanical or metallurgical nature. Figure 4 shows the fractography of the diamond/Al composites under three-point bending testing, both with and without Ti coating. The plastically deformed Al matrix exhibited network-like dimples on the fracture surface, which is a typical ductile behavior of pure metals. In the area denoted by the green arrows, part of the diamond particles were unwrapped by the Al matrix, indicating a weak interfacial bonding strength between the diamond and Al. However, in the area specified by the red arrows, the Al of the composite was not fully covered with the diamond particles, resulting in the formation of a gap. The gap, i.e., the arch bridge phenomenon, hinders the transmission of phonons at the interface between the matrix and reinforcement, reducing the TC of the composite [4,28]. By comparing Figure 4a,b, it can be seen that the Ti coating on the diamond surface can effectively reduce the phenomenon of unwrapping and gaps, increasing the interface bonding strength of the diamond/Al composite, which also improves the relative density of the composite. Ti coating can improve the interface bonding state of diamond/Al composites, which is consistent with the results of fractography of composites prepared by an alternative method [25]. High relative density is a prerequisite for obtaining diamond/Al composite materials with high TC [13,16].

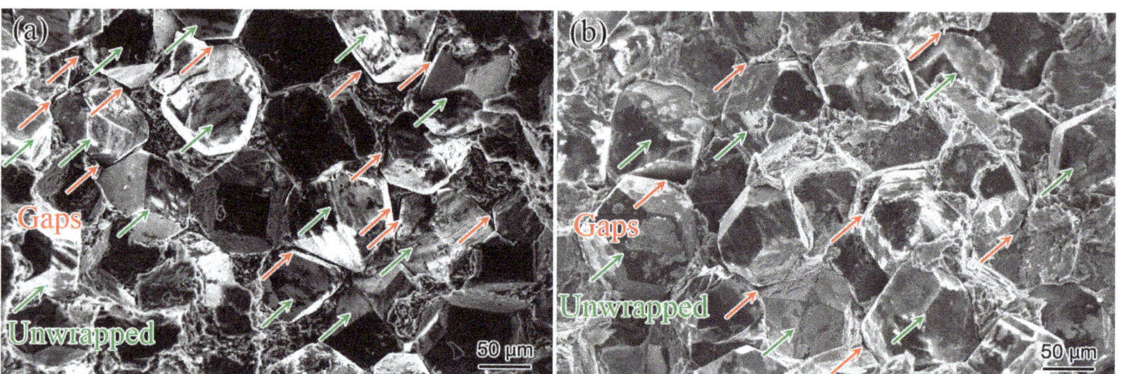

Figure 4. Fractography of the diamond/Al composites: (**a**) without Ti coating; (**b**) with Ti coating.

3.3. Interfacial Characteristics of Ti-Diamond/Al Composites

Figure 5 shows the distribution of interfacial elements in the diamond/Al composite with Ti-coated diamond particles using EMPA. Due to the limited solid solubility of diamond and Al to dissolve into each other, an interdiffusion area of approximately 8 μm was generated, as shown in the red line area in Figure 5b. The Ti coating on the surface of diamond particles diffused toward the Al side, demonstrating that metallurgical bonding has been formed in the interface region between diamond and Al, which enhances the performance of the composites [21,23].

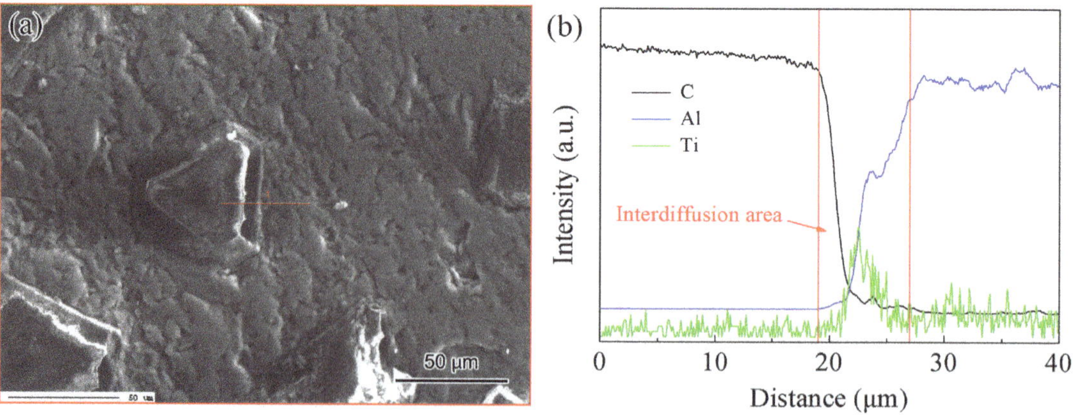

Figure 5. Interface element distribution of the diamond/Al composite with Ti-coated diamond particles: (**a**) SEM image; (**b**) EMPA map.

Figure 6 manifests the XRD patterns of the diamond/Al composite with Ti-coated diamond particles. The patterns were obtained in different diffraction angle ranges at a scanning speed of 4 and 0.25°/min. Figure 6a demonstrates the formation of Al$_3$Ti, Al$_2$Ti, Al$_5$Ti$_3$, and Ti$_9$Al$_{23}$ intermetallic compounds in the diamond/Al composite with Ti-coated diamond particles, which results from the long-term reaction between Ti coating and Al matrix at the fabrication temperature during the LSS process. The intermetallic compounds tightly connect the Al matrix with diamond particles, improving the bonding strength between the Al and diamond phases [21,25]. Figure 6b indicates that the Al$_4$C$_3$ phase was not detected even at the scanning speed of 0.25°/min. Additionally, a trace peak of Al$_2$O$_3$ was discovered, which can be attributed to the oxidation of the composite during the storage. A continuous and well-bonded interfacial structure without the Al$_4$C$_3$ phase is the key to improving the TC and stability of diamond/Al composites [18]. The EMPA and XRD confirm that Ti coating on the diamond surface is conducive to forming a well-bonded interface.

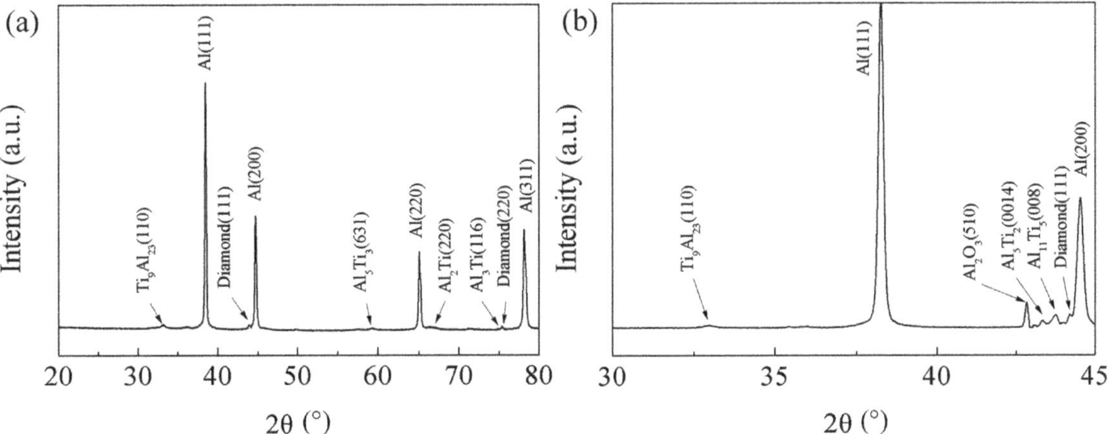

Figure 6. XRD patterns of the diamond/Al composite with Ti-coated diamond particles in different diffraction angle ranges: (**a**) between 20° and 80°; (**b**) between 30° and 45°.

3.4. Bending Strength of Diamond/Al Composites

Figure 7 compares the relative density and bending strength of the diamond/Al composites fabricated by the LSS process in this study and those reported in the literature [21]. Compared with uncoated diamond/Al composites, the relative density of Ti-diamond/Al composites increased from 97.22% to 98.08% with a growth rate of 0.88%, and the bending strength increased from 97.46 MPa to 142.54 MPa with a growth rate of 46.25%. Meanwhile, the bending fractography in Figure 4 shows that a more interfacial gap is observed between the uncoated diamond and the Al matrix. The separation occurs due to the significant disparity in the CTE between diamond (1.0×10^{-6}/K) and Al (23.0×10^{-6}/K) when undergoing the cooling process. Therefore, the low relative density of uncoated diamond/Al composites is attributed to the abundance of gaps surrounding the diamond particles. In other words, the Ti coating plays a vital role in promoting interfacial bonding and improving mechanical properties. Furthermore, certain studies have demonstrated that a thick Ti coating can effectively establish a strong bond between the diamond particles and the Al matrix, hence enhancing the composite interface [25]. By comparing Figure 7, it can also be seen that in the diamond/Al composites prepared by the LSS technology, the bending strength and relative density of the Ti coating on the diamond surface are better than those of the Cr coating. The CTE of Ti is 10.8×10^{-6}/K, whereas that of Cr is 6.2×10^{-6}/K. Compared with the Cr coating, the Ti coating, as an intermediate layer, can better buffer the difference in CTE between the diamond and Al matrix, further reducing gaps and increasing the relative density of the diamond/Al composite. Therefore, Ti-coated diamond/Al composites have superior bending strength compared to Cr-coated diamond/Al composites.

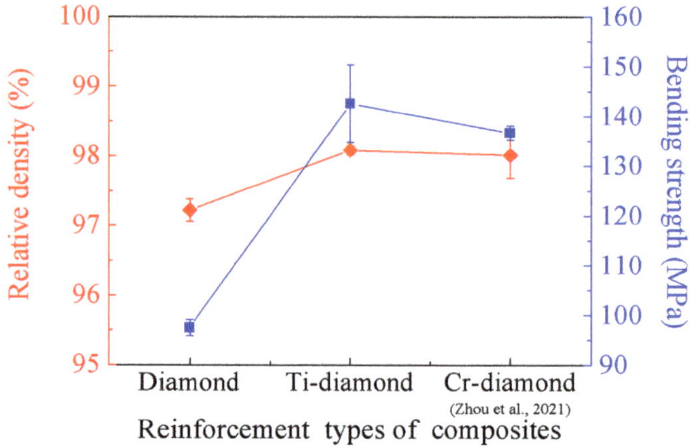

Figure 7. Variation in the relative density and bending strength of diamond/Al composites fabricated by the LSS process [21].

3.5. Thermal Conductivity of Diamond/Al Composites

Figure 8 illustrates the TC of the diamond/Al composite both without and with Ti coating. The lack of wetting capacity between the diamond and the Al matrix results in a TC of only 149 W/m·K in uncoated diamond particle-reinforced Al matrix composites. In contrast, the TC of Ti-coated diamond/Al composites exhibited a significant increase of 85.9%, reaching a value of 277 W/m·K. Ti coating can significantly improve the interfacial bonding between diamond and Al, leading to an increase in the TC of composites, which is consistent with other study results [24]. It is worth noting that the TC of the Ti-diamond/Al composite is lower than that of the Cr-diamond/Al composite prepared by the same method [21]. This difference in TC could be the reason why the interfacial products of the Ti-diamond/Al composite are more intricate, as evidenced by the detection of intermetallic compounds using X-ray diffraction (XRD) in Figure 6. The complex intermetallic compounds incorporated extra interface layers that act as thermal boundary barriers. So, the interfacial thermal conductance (ITC) and TC of Ti-coated diamond/Al composites decrease significantly with the increase in the thickness of the intermetallic layer, which is due to the very low TC of the intermetallic layer generated by Ti and Al [26]. The thickness of the intermetallic layer reached 8 μm in this study, as shown in the red line area in Figure 5b. In addition, Cr coating has a greater sound velocity compared to Ti coating, which is also the reason for the higher TC of its composite [26]. The interfacial bond between diamond and Al can be improved by the appropriate thickness of the intermetallic layer, which enhances the TC of the composite. However, an excessively thick intermetallic layer with low TC increases the interfacial thermal resistance and thus reduces the TC of the composite [29]. Therefore, it is imperative to meticulously regulate the thickness of the intermetallic layer.

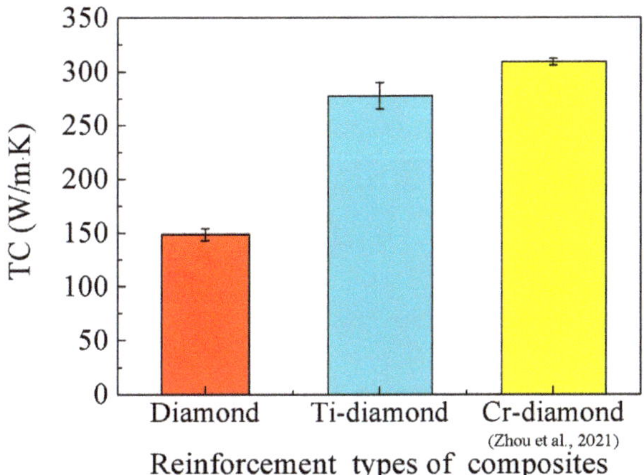

Figure 8. TC of the diamond/Al composite fabricated by the LSS process [21].

3.6. Theoretical Thermal Conductivity of Diamond/Al Composites

The heat transfer between metal and non-metal phases depends on the coupling effect of electrons (on the metal side) and phonons (on the non-metal side) [30]. The presence of impurities, holes, and other defects at the interface between the two phases would cause phonon scattering, leading to a reduction in the mean free path of phonons and thus deteriorating the TC of composites [4]. The DEM (differential effective medium) [31] and Maxwell [32] are widely used in the TC prediction for the diamond/Al composites by comprehensively considering many factors, such as ITC and acoustic mismatch theory [33]. These models can be expressed as

$$\left(\frac{TC}{TC_m}\right)^{\frac{1}{3}}(1-V_d) = \frac{\frac{TC_d^{eff}}{TC_m} - \frac{TC}{TC_m}}{\frac{TC_d^{eff}}{TC_m} - 1}, \tag{1}$$

$$TC = \frac{TC_m \times [2 \times TC_m + TC_d^{eff} + 2 \times (TC_d^{eff} - TC_m) \times V_d]}{2 \times TC_m + TC_d^{eff} - (TC_d^{eff} - TC_m) \times V_d}, \tag{2}$$

where TC_m and TC_d^{eff} are the TC of the Al and the effective TC of the diamond, respectively, and V_d is the volume fraction of diamond particles. Considering the influence of ITC and diamond particle size on the TC of reinforcement, TC_d^{eff} can be calculated as

$$TC_d^{eff} = \frac{TC_d^{in}}{1 + \frac{TC_d^{in}}{r \times ITC}}, \tag{3}$$

where r and TC_d^{in} are the average radius and TC of single-crystal diamond particles, respectively. The TC_d^{in} of the synthetic single-crystal diamond has a linear relationship with its N element content and decreases with the increase in N content ($[N]$) [34,35]. According to the equation (i.e., $[N] = 5.5 \times 25 \times A_{1130}/A_{2120}$), the N content can be estimated from the intensity ratio of the relative absorption coefficient (A) [36]. Figure 9 shows the intensity of the absorption peak at wavenumbers of 1130 and 2120/cm. The N content of diamond particles used in this experiment was about 330 ppm, and the TC of diamond was about 1121 W/m·K, according to the equation ($TC_d^{in} = 2200 - 3.27[N]$) [34].

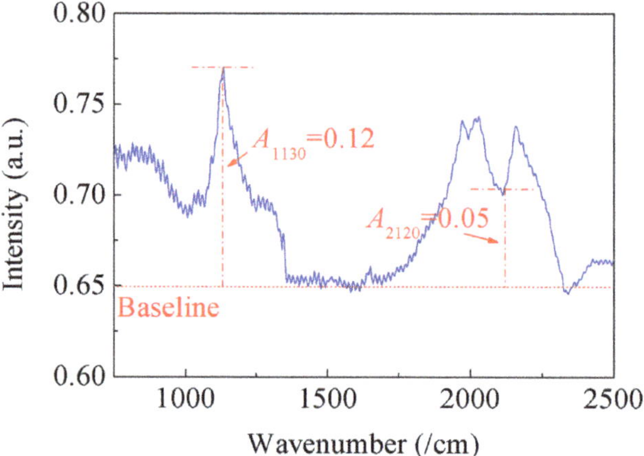

Figure 9. FTIR diffuse reflectance spectrum of the diamond particle.

Then, the ITC between the Al matrix and diamond particles can be described as

$$\text{ITC} = \frac{1}{4} \times \rho_\text{m} \times c_{pm} \times C_\text{Dm} \times \eta_{1-2},\tag{4}$$

where ρ_m, c_{pm}, and C_Dm are the mass density, specific heat capacity, and sound velocity for the matrix. Then, η_{1-2} is the average transmission coefficient of phonons across the interface from Al to diamond, which can be denoted as

$$\eta_{1-2} = \frac{2 \times Z_\text{m} \times Z_\text{d}}{(Z_\text{m} + Z_\text{d})^2} \times \left(\frac{C_\text{Dm}}{C_\text{Dd}}\right),\tag{5}$$

where Z_m and Z_d are the phonon impedance (according to the equation: $Z = \rho C_\text{D}$) for the matrix and reinforcement, respectively, and C_Dd is the sound velocity for the reinforcement. The sound velocity of the matrix and the reinforcement can be calculated according to the Debye sound velocity (C_D), and the theoretical calculation can be established as

$$C_\text{D} = \frac{1}{\sqrt{\frac{1}{2} \times \left(\frac{1}{C_l^2} + \frac{1}{C_t^2}\right)}},\tag{6}$$

where C_l and C_t are longitudinal phonon velocity and transversal phonon velocity, respectively. Table 1 represents the physical parameters of raw materials for calculation in this study. Inputting the data into Equations (1) and (2) calculates the theoretical TC of DEM and Maxwell as 389 and 382 W/m·K, respectively, and the measured values reach 71.2 and 72.5% of the theoretical values, respectively. The effect of coating on interfacial bonding leads to the improvement of the TC of the composite. However, the TC of the composite is still far from the theoretical models of DEM and Maxwell due to the intrinsic interfacial thermal resistance of the coating.

Table 1. Physical parameters of raw materials [32,37,38].

Material	Phonon Velocity		C_D (m/s)	η_{1-2}	ITC (W/m²·K)	c_p (J/g·K)
	C_l (m/s)	C_t (m/s)				
Al	6240	3040	3865			0.895
Diamond	20,000	12,300	14,817	0.019	4.43×10^7	0.500

Possible factors contributing to the substantial discrepancy between the actual measured values and the calculated values of thermal conductivity are as follows: (i) The theoretical model is based on the premise that diamond particles are spherical, while in reality, high-quality diamond particles are hexoctahedrons, and the shape factor will reduce the interfacial heat transfer coefficient. (ii) The effect of Ti coating on diffusion rate is neglected in the calculation of ITC. (iii) The intermetallic compounds formed at the interface with low intrinsic TC deteriorate the interfacial thermal resistance of the composite. (iv) The theoretical model does not consider the effect of internal defects on the TC of the composite, and internal defects are the inevitable structure of the composite.

4. Future Research Directions

The design of the thermal conduction path at the interface between matrix and reinforcement will directly affect the TC property of the diamond/Al composites. The metal matrix alloying method has been proven to be able to utilize the diffusion dynamics of matrix alloy elements to build parallel structural carbides, reducing the interface thermal resistance [39]. However, the alloying elements in the matrix act as additional interface thermal resistance, which offsets the effect of part of the parallel structure in reducing the total thermal resistance. Diamond coating is the only second interface modification method in this field. How to synthesize discontinuous metal coating in situ on the diamond surface and then prepare composites with parallel structure thermal conduction paths is the future development direction.

5. Conclusions

The independently developed liquid–solid separation (LSS) technology has been applied to prepare 40 vol.% diamond particle-reinforced diamond/Al composites. The interfacial characteristics, bending strength, and thermal conductivity (TC) of the diamond/Al composites without and with a thickness of 100 nm of Ti coating were evaluated comprehensively. The main conclusions can be summarized as follows:

(1) The LSS technology, characterized by its low heating temperature and short holding time, prevents the formation of Al_4C_3 at the interface of diamond/Al composites, hence preserving their mechanical and thermophysical properties.

(2) The inclusion of Ti coating formed Al_3Ti, Al_2Ti, Al_5Ti_3, and Ti_9Al_{23} intermetallic compounds, resulting in metallurgical bonding between diamond and Al, improving the interfacial bonding strength and TC of the diamond/Al composites.

(3) The TC increased from 149 W/m·K for the diamond/Al composite to 277 W/m·K for the diamond/Al composite with Ti-coated diamond particles, with a growth rate of 85.9%.

(4) A comparison of the present Ti coating composite with previously reported Cr coating composite was performed on the features. The application of the Ti coating can increase the bending strength of the composite, while the Cr coating can enhance the TC of the composite.

(5) The Ti coating promotes interfacial bonding but also introduces extra interfacial thermal resistance. Combined with the idealization of the model design, results for TC reached only 71.2% with the DEM model and 72.5% with the Maxwell model, respectively.

Author Contributions: Conceptualization, H.Z.; methodology, H.Z., Q.J., J.S. and Y.H.; validation, H.Z., Q.J., W.B. and W.Z.; investigation, H.Z., Y.L., Y.H. and W.B.; data curation, Q.J.; formal analysis, Q.J. and Y.H.; visualization, J.S. and Y.L.; resources, J.S. and W.B.; writing—original draft preparation, H.Z.; writing—review and editing, W.Z.; project administration, H.Z., Y.L. and W.Z.; funding acquisition, H.Z.; supervision, W.Z. All authors have read and agreed to the published version of the manuscript.

Funding: This research was funded by the National Natural Science Foundation of China, grant number 52301030 and the Proof of Concept Project for the University Alliance in Shahe Higher Education Park, grant number SHGNYZ202304. The APC was funded by the National Natural Science Foundation of China.

Informed Consent Statement: Not applicable.

Data Availability Statement: The datasets used and analyzed during the current study are available from the corresponding author upon reasonable request.

Conflicts of Interest: Author Qijin Jia was employed by the company Beijing System Design Institute of Electro-Mechanic Engineering. Author Jing Sun was employed by the company Beijing Hangxing Machinery Co., Ltd. The remaining authors declare that the research was conducted in the absence of any commercial or financial relationships that could be construed as a potential conflict of interest.

References

1. Suh, D.; Moon, C.M.; Kim, D.; Baik, S. Ultrahigh thermal conductivity of interface materials by silver-functionalized carbon nanotube phonon conduits. *Adv. Mater.* **2016**, *28*, 7220–7227. [CrossRef]
2. Francis, D.; Kuball, M. GaN-on-diamond materials and device technology: A review. In *Thermal Management of Gallium Nitride Electronics*; Tadjer, M.J., Anderson, T.J., Eds.; Woodhead Publishing: Sawston, UK, 2022; pp. 295–331.
3. Abdallah, Z.; Pomeroy, J.W.; Neubauer, E.; Kuball, M. Thermal characterization of metal-diamond composite heat spreaders using low-frequency-domain thermoreflectance. *ACS Appl. Electron. Mater.* **2023**, *5*, 5017–5024. [CrossRef] [PubMed]
4. Liu, X.Y.; Wang, W.G.; Wang, D.; Ni, D.R.; Chen, L.Q.; Ma, Z.Y. Effect of nanometer TiC coated diamond on the strength and thermal conductivity of diamond/Al composites. *Mater. Chem. Phys.* **2016**, *182*, 256–262. [CrossRef]
5. Matthey, B.; Kunze, S.; Kaiser, A.; Herrmann, M. Thermal properties of SiC-bonded diamond materials produced by liquid silicon infiltration. *Open Ceram.* **2023**, *15*, 100386. [CrossRef]
6. Godbole, K.; Bhushan, B.; Narayana Murty, S.V.S.; Mondal, K. Al-Si controlled expansion alloys for electronic packaging applications. *Prog. Mater. Sci.* **2024**, 101268. [CrossRef]
7. Zweben, C. Advances in composite materials for thermal management in electronic packaging. *JOM* **1998**, *50*, 47–51. [CrossRef]
8. Jacobson, P.; Stoupin, S. Thermal expansion coefficient of diamond in a wide temperature range. *Diam. Relat. Mater.* **2019**, *97*, 107469. [CrossRef]
9. Kondakci, E.; Solak, N. Enhanced thermal conductivity and long-term stability of diamond/aluminum composites using SiC-coated diamond particles. *J. Mater. Sci.* **2022**, *57*, 3430–3440. [CrossRef]
10. Chak, V.; Chattopadhyay, H.; Dora, T.L. A review on fabrication methods, reinforcements and mechanical properties of aluminum matrix composites. *J. Manuf. Process.* **2020**, *56*, 1059–1074. [CrossRef]
11. Anisimova, M.; Knyazeva, A.; Sevostianov, I. Effective thermal properties of an aluminum matrix composite with coated diamond inhomogeneities. *Int. J. Eng. Sci.* **2016**, *106*, 142–154. [CrossRef]
12. Kidalov, S.V.; Shakhov, F.M. Thermal conductivity of diamond composites. *Materials* **2009**, *2*, 2467–2495. [CrossRef]
13. Kwon, H.; Leparoux, M.; Heintz, J.-M.; Silvain, J.-F.; Kawasaki, A. Fabrication of single crystalline diamond reinforced aluminum matrix composite by powder metallurgy route. *Met. Mater. Int.* **2011**, *17*, 755–763. [CrossRef]
14. Khalid, F.A.; Beffort, O.; Klotz, U.E.; Keller, B.A.; Gasser, P. Microstructure and interfacial characteristics of aluminium-diamond composite materials. *Diam. Relat. Mater.* **2004**, *13*, 393–400. [CrossRef]
15. Ji, G.; Tan, Z.Q.; Lu, Y.G.; Schryvers, D.; Li, Z.Q.; Zhang, D. Heterogeneous interfacial chemical nature and bonds in a W-coated diamond/Al composite. *Mater. Charact.* **2016**, *112*, 129–133. [CrossRef]
16. Tan, Z.; Ji, G.; Addad, A.; Li, Z.; Silvain, J.-F.; Zhang, D. Tailoring interfacial bonding states of highly thermal performance diamond/Al composites: Spark plasma sintering vs. vacuum hot pressing. *Compos. Part A Appl. Sci.* **2016**, *91*, 9–19. [CrossRef]
17. Monje, I.E.; Louis, E.; Molina, J.M. Role of Al_4C_3 on the stability of the thermal conductivity of Al/diamond composites subjected to constant or oscillating temperature in a humid environment. *J. Mater. Sci.* **2016**, *51*, 8027–8036. [CrossRef]
18. Xin, L.; Xing, T.; Shu, Y.W.; Qin, C.G.; Jing, Q.; Jiao, H.F.; Qiang, Z.; Hui, W.G. Enhanced stability of the Diamond/Al composites by W coatings prepared by the magnetron sputtering method. *J. Alloys Compd.* **2018**, *763*, 305–313. [CrossRef]
19. Zhou, H.Y.; Yin, Y.L.; Wu, C.J.; Liu, J.Y. Microstructures and properties of diamond/Al composites prepared by liquid-solid separation technology. *Chin. J. Nonferrous Met.* **2017**, *27*, 1855–1861.
20. Zhou, H.Y.; Yin, Y.L.; Shi, Z.L.; Wu, C.J.; Liu, J.Y. The fabrication of Al-diamond composites for heat dissipation by liquid–solid separation technology. *J. Mater. Sci. Mater. Electron.* **2017**, *28*, 721–728. [CrossRef]
21. Zhou, H.Y.; Ran, M.R.; Li, Y.Q.; Yin, Z.; Tang, Y.H.; Zhang, W.D.; Zheng, W.Y.; Liu, J.Y. Improvement of thermal conductivity of diamond/Al composites by optimization of liquid-solid separation process. *J. Mater. Process. Technol.* **2021**, *297*, 117267. [CrossRef]
22. Zhou, H.Y.; Li, Y.Q.; Wang, H.M.; Ran, M.R.; Tong, Z.; Zhang, W.D.; Liu, J.Y.; Zheng, W.Y. Fabrication of functionally graded diamond/Al composites by liquid–solid separation technology. *Materials* **2021**, *14*, 3205. [CrossRef]

23. Kusuadi, N.I.N.; Jamal, N.A.; Ahmad, Y. A Short review on diamond reinforced aluminium composites. In Proceedings of the 5th International Conference on Advances in Manufacturing and Materials Engineering, Singapore, 15–17 December 2023; Maleque, M.A., Ahmad Azhar, A.Z., Sarifuddin, N., Syed Shaharuddin, S.I., Mohd Ali, A., Abdul Halim, N.F.H., Eds.; Springer Nature: Singapore, 2023; pp. 55–61.

24. Che, Z.F.; Wang, Q.X.; Wang, L.H.; Li, J.W.; Zhang, H.L.; Zhang, Y.; Wang, X.T.; Wang, J.G.; Kim, M.J. Interfacial structure evolution of Ti-coated diamond particle reinforced Al matrix composite produced by gas pressure infiltration. *Compos. Part B Eng.* **2017**, *113*, 285–290. [CrossRef]

25. Zhang, H.L.; Wu, J.H.; Zhang, Y.; Li, J.W.; Wang, X.T.; Sun, Y.H. Mechanical properties of diamond/Al composites with Ti-coated diamond particles produced by gas-assisted pressure infiltration. *Mater. Sci. Eng. A* **2015**, *626*, 362–368. [CrossRef]

26. Tan, Z.Q.; Li, Z.Q.; Xiong, D.B.; Fan, G.L.; Ji, G.; Zhang, D. A predictive model for interfacial thermal conductance in surface metallized diamond aluminum matrix composites. *Mater. Des.* **2014**, *55*, 257–262. [CrossRef]

27. Molina-Jordá, J.M. Thermal conductivity of metal matrix composites with coated inclusions: A new modelling approach for interface engineering design in thermal management. *J. Alloys Compd.* **2018**, *745*, 849–855. [CrossRef]

28. Mizuuchi, K.; Inoue, K.; Agari, Y.; Sugioka, M.; Tanaka, M.; Takeuchi, T.; Tani, J.-i.; Kawahara, M.; Makino, Y.; Ito, M. Bimodal and monomodal diamond particle effect on the thermal properties of diamond-particle-dispersed Al–matrix composite fabricated by SPS. *Microelectron. Reliab.* **2014**, *54*, 2463–2470. [CrossRef]

29. Lakra, S.; Bandyopadhyay, T.K.; Das, S.; Das, K. Thermal conductivity of in-situ dual matrix aluminum composites with segregated morphology. *Mater. Res. Bull.* **2021**, *144*, 111515. [CrossRef]

30. Mozafarifard, M.; Liao, Y.L.; Nian, Q.; Wang, Y. Two-temperature time-fractional model for electron-phonon coupled interfacial thermal transport. *Int. J. Heat Mass Transfer* **2023**, *202*, 123759. [CrossRef]

31. Tavangar, R.; Molina, J.M.; Weber, L. Assessing predictive schemes for thermal conductivity against diamond-reinforced silver matrix composites at intermediate phase contrast. *Scr. Mater.* **2007**, *56*, 357–360. [CrossRef]

32. Maxwell, J.C. *A Treatise on Electricity and Magnetism*; Oxford University Press: Oxford, UK, 1998.

33. Hua, Z.L.; Wang, K.; Li, W.F.; Chen, Z.Y. Theoretical strategy for interface design and thermal rerformance prediction in diamond/aluminum composite based on scattering-mediated acoustic mismatch model. *Materials* **2023**, *16*, 4208. [CrossRef]

34. Yamamoto, Y.; Imai, T.; Tanabe, K.; Tsuno, T.; Kumazawa, Y.; Fujimori, N. The measurement of thermal properties of diamond. *Diam. Relat. Mater.* **1997**, *6*, 1057–1061. [CrossRef]

35. Monje, I.E.; Louis, E.; Molina, J.M. On critical aspects of infiltrated Al/diamond composites for thermal management: Diamond quality versus processing conditions. *Compos. Part A Appl. Sci.* **2014**, *67*, 70–76. [CrossRef]

36. Abyzov, A.M.; Kruszewski, M.J.; Ciupiński, Ł.; Mazurkiewicz, M.; Michalski, A.; Kurzydłowski, K.J. Diamond–tungsten based coating–copper composites with high thermal conductivity produced by Pulse Plasma Sintering. *Mater. Des.* **2015**, *76*, 97–109. [CrossRef]

37. Kida, M.; Weber, L.; Monachon, C.; Mortensen, A. Thermal conductivity and interfacial conductance of AlN particle reinforced metal matrix composites. *J. Appl. Phys.* **2011**, *109*, 064907. [CrossRef]

38. Hasselman, D.P.H. Thermal diffusivity and conductivity of composites with interfacial thermal contact resistance. In *Thermal Conductivity 20*; Hasselman, D.P.H., Thomas, J.R., Eds.; Springer: Boston, MA, USA, 1989; pp. 405–413.

39. Bai, G.Z.; Wang, L.H.; Zhang, Y.J.; Wang, X.T.; Wang, J.G.; Kim, M.J.; Zhang, H.L. Tailoring interface structure and enhancing thermal conductivity of Cu/diamond composites by alloying boron to the Cu matrix. *Mater. Charact.* **2019**, *152*, 265–275. [CrossRef]

Article

Analyses and Research on a Model for Effective Thermal Conductivity of Laser-Clad Composite Materials

Yuedan Li [1], Chaosen Lin [1], Bryan Gilbert Murengami [2], Cuiyong Tang [1] and Xueyong Chen [1,*]

[1] College of Mechanical and Electrical Engineering, Fujian Agriculture and Forestry University, Fuzhou 350001, China
[2] College of Mechanical and Electronic Engineering, Northwest A&F University, Yangling, Xianyang 712100, China
* Correspondence: xueyongchen@fafu.edu.cn

Abstract: Composite materials prepared via laser cladding technology are widely used in die production and other fields. When a composite material is used for heat dissipation and heat transfer, thermal conductivity becomes an important parameter. However, obtaining effective thermal conductivity of composite materials prepared via laser cladding under different parameters requires a large number of samples and experiments. In order to improve the research efficiency of thermal conductivity of composite materials, a mathematical model of Cu/Ni composite materials was established to study the influence of cladding-layer parameters on the effective thermal conductivity of composite materials. The comparison between the model and the experiment shows that the model's accuracy is 86.7%, and the error is due to the increase in thermal conductivity caused by the alloying of the joint, so the overall effective thermal conductivity deviation is small. This study provides a mathematical model method for studying the thermodynamic properties of laser cladding materials. It provides theoretical and practical guidance for subsequent research on the thermodynamic properties of materials during die production.

Keywords: mathematical modeling; thermal conductivity; heat transfer material; laser cladding

Citation: Li, Y.; Lin, C.; Murengami, B.G.; Tang, C.; Chen, X. Analyses and Research on a Model for Effective Thermal Conductivity of Laser-Clad Composite Materials. *Materials* **2023**, *16*, 7360. https://doi.org/10.3390/ma16237360

Academic Editor: Antonio Riveiro

Received: 18 October 2023
Revised: 17 November 2023
Accepted: 20 November 2023
Published: 27 November 2023

1. Introduction

The utilization of multiple material structures to create composite materials with diverse material properties has proven to be a valuable approach for enhancing material characteristics and tailoring them to specific requirements [1]. These solutions offer numerous advantages and have found widespread applications in the aerospace industry [2]. Laser cladding is a technology in which the metal surface is melted by using a high-energy laser beam to form a molten pool, and the metal powder is deposited on the metal surface under the action of laser energy. The resulting clad layer exhibits unique properties influenced by various parameters, including laser power, scanning speed, and cladding process parameters [3–8].

Laser cladding can produce high-hardness and wear-resistant coatings with long-term stability [9–11]. Research shows that laser cladding technology can improve the service of coatings in high-temperature and other harsh environments [12]. This method has great value and significance for manufacturing reinforced coatings on the surface of high-thermal-conductivity materials.

A significant advantage of laser cladding lies in its ability to rapidly produce cladding layers with varying thicknesses and compositions. However, determining the effective thermal conductivity under different material thicknesses necessitates a substantial number of samples and experiments [13]. While past research on laser cladding has primarily focused on enhancing mechanical and wear properties, the evaluation and prediction of thermal conductivity in laser-clad composites has received relatively limited attention [14]. Reliable prediction of effective thermal conductivity is an important aspect of material design strategy for key engineering applications [15,16]. Therefore, establishing a thermal resistance digital model of composites fabricated via laser cladding will become a preliminary but key step for the rapid selection of cladding-layer materials and process parameters to achieve efficient composite design in the future.

The existing literature shows that the thermal conductivity models of composites can be roughly divided into three types: empirical models, finite element models, and theoretical models. Empirical models are mostly applied to mixed materials that cannot be combined into compounds. Yang et al. [17] prepared different test blocks of rubber sand composites through different proportions and particle sizes for multiple series of thermal conductivity tests, normalized them, and established a prediction model. The prediction model established by this method had good accuracy. Finite element models are mostly used in composites with complex three-dimensional structure models. Yan et al. [18] built a model of uranium dioxide (UO_2)/silicon carbide (SIC) composite by connecting the grains of the two materials with different proportions randomly generated by a computer, and calculated and analyzed its effective thermal conductivity using the finite element method (FEM). The model managed to reduce the design speed and costs. Most of the theoretical models are applied to three-dimensional structural composites with certain rules. Fang He et al. [16] proposed a thermal conductivity model of aerogel-filled thermal insulation composites. According to the dispersion of aerogel particles, the discrete model of aerogel composites was established to generate gaps and holes, and the corresponding thermal conductivity was derived. In a study by Witold weglewski et al. [19], the relationship between a coating material and its thermal conductivity was modeled using the finite element method (FEM) and variational asymptotic method (VAM). Using micro-CT modeling technology, the microstructure of the composite was used to generate the finite element mesh model, and the thermal conductivity of the composite was calculated according to the model results.

Compared with theoretical models, it can be seen that empirical models and finite element models have strong generality for the prediction of the thermal conductivity of composite materials. The disadvantage is that when used in similar composite materials, these models need to be reestablished according to the material properties. A finite element method model is slow to establish, and the modeling time is long; the universality of empirical formula is poor, and repeated tests are needed for different materials.

Considering the above problems, laser-clad composites have structural solid similarities. This project aims to develop a theoretical model to rapidly predict the effective thermal conductivity of copper/nickel gradient materials. The model considers the thermal conductivity of dissimilar materials and the cladding layer's thickness. The effects of different parameters on the thermal conductivity of composites were verified by means of model prediction and experimental comparison. Using the established general model, the influence of key parameters on the effective thermal conductivity of composites prepared via laser cladding was studied, which has theoretical significance for guiding the parameter design of laser-clad composites, improving the design speed, and reducing the development cost.

2. Physical Model of Composites

The structure of laser-clad composite materials has a certain degree of regularity. To accurately predict the effective thermal conductivity of composite materials, their structure is examined. The light source energy of laser cladding is Gaussian distributed [20], which is explained by the Gaussian heat source model shown in Figure 1 [21]. In this model, the heat source energy is distributed in a circular plane according to a Gaussian function, and calculated by using Equation (1):

$$q(x, y) = \frac{\eta Q}{2\pi r^2} exp(\frac{x^2 + y^2}{2r^2})$$
(1)

where $q(x, y)$ is the density of heat flux at the coordinate point; η is the rate of laser energy absorption by the surface of the material; Q is the laser light source's output; and r is the radius length of the laser source.

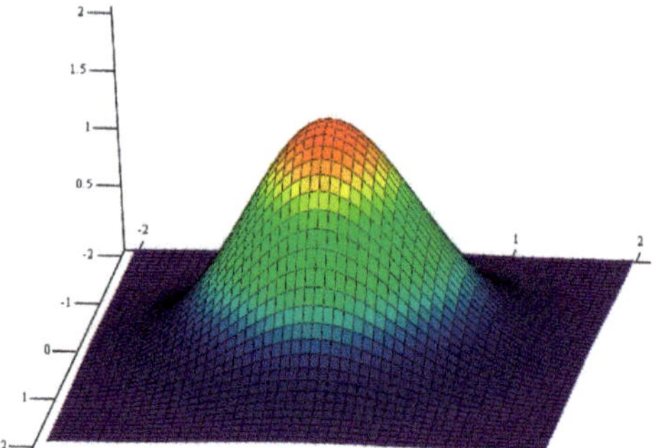

Figure 1. Gaussian-distribution diagram.

The surface of metal materials absorbs energy to melt and combine with the metal in a short period of time. Under the action of a Gaussian-distributed laser light source, the surface area of the joint after laser cladding can be approximated as a Gaussian distribution. As shown in Figure 2, the periodic structure generated plays a crucial role in the heat conduction of composite materials. A periodic composite material structure model was proposed for the convenience of calculation to predict the effective thermal conductivity of composite materials.

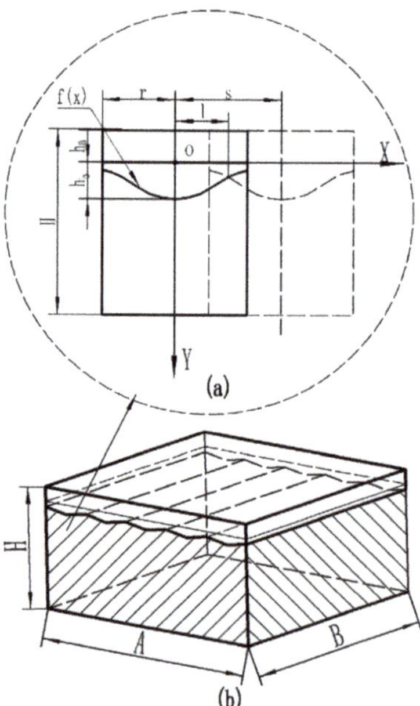

Figure 2. Physical model of composites. (**a**) Composite material unit model. (**b**) The overall distribution model of composite materials.

In this figure, A represents the length of the composite material; B represents its width; H represents the composite element model's total height; S represents the distance between adjacent laser claddings; R is the radius of the laser's light source cladding; h_a

is the thickness of the cladding layer's material; h_b is the height of the first laser cladding layer; and (b) is the composite element model.

The uppermost layer comprises the surface cladding material, while the intermediate layer is the bonding layer. The function of the cladding material's connection to the substrate is the Gaussian distribution of the formula $f(x) = h_b exp\left(-x^2/(0.5r)^2\right)$. The overall distribution model of composite materials is (b).

3. Effective Thermal Conductivity Model of Composites

A thermal conductivity model of composite materials prepared via laser cladding is investigated in this paper. The principal research factors are defined as the cladding-layer characteristics, process parameters, and structural characteristics. Considering the model's precision and the need to reduce the calculation's complexity, the following parameters are simplified:

1. The interior of metal is uniform and devoid of cavities.
2. During laser cladding, the alloying of two dissimilar alloy materials is disregarded.
3. The material's surface is polished and evaluated after preparation; therefore, the oxide layer is not considered, and the thermal resistance of the interface is not accounted for in the calculation.
4. The element model is a cross-section perpendicular to the direction of laser cladding.

3.1. Model of Element

Figure 3 depicts the composite element model clad in laser; s is the distance between two element models, and they overlap. The composite material consists of three distinct components. The first component is the cladding layer, the second is the cladding layer's bonding position, and the third is the base material. Under the same material, it is presumed that temperature is linearly distributed in a single effective conduction direction [16]. The effective actual cell width is determined as follows:

$$L = 2l = 2r - s \tag{2}$$

where r is the radius of the laser cladding, s is the distance between adjacent laser claddings, and l is half the effective unit length. At the intersection, according to the previously established boundary approximation curve $f(x)$,

$$f(x) = h_b \cdot e^{\frac{-x^2}{(0.5r)^2}} \tag{3}$$

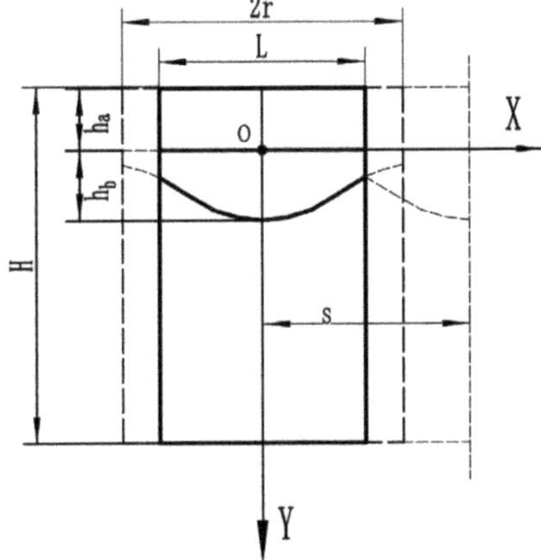

Figure 3. Composite material unit model.

Since the model's contact surface is a Gaussian function, the joint surfaces are non-parallel. To accomplish the purpose of calculating the equivalent thermal calculation, it is subdivided vertically into a number of small ideal elements. When the difference is negligible, it may be considered parallel. Figure 4 depicts this. In this investigation, it is divided symmetrically along the Y axis into two regions that are considered to have parallel thermal resistance. The final result will incline toward a value as the regional mean fraction P increases.

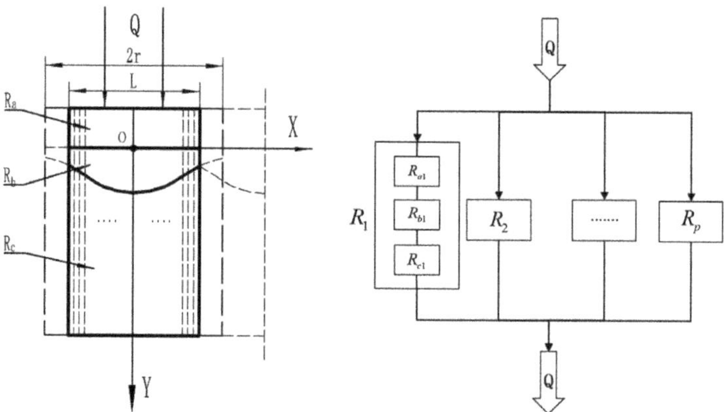

Figure 4. Unit model is divided into multiple parallel thermal resistance regions.

According to the thermal resistance calculation formula, the equivalent thermal resistance of the unit model is calculated as follows:

$$\frac{1}{R_o} = 2 \cdot \left(\frac{1}{R_1} + \frac{1}{R_2} + \cdots\cdots + \frac{1}{R_{p-1}} + \frac{1}{R_p} \right) \tag{4}$$

R_o is the total equivalent thermal resistance of the unit, and Figure 5 illustrates the small ideal unit model after equalization.

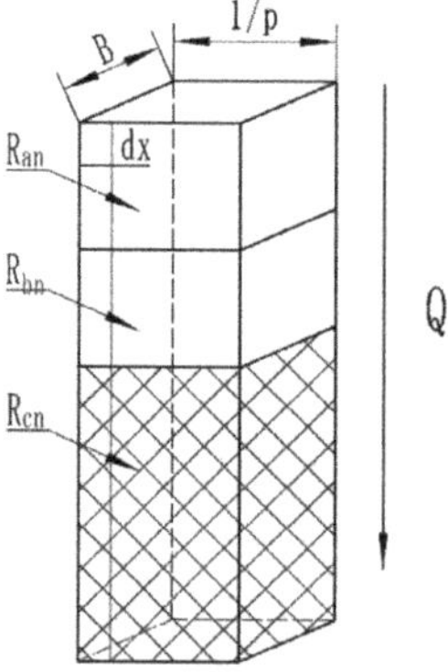

Figure 5. Small ideal unit model.

The total thermal resistance of small ideal units at various positions is assumed to be R_n; $R_{an}R_{bn}R_{cn}$ is connected in series in accordance with the thermal resistance at various positions. The total thermal resistance R_n of the small ideal units is equal to the sum of the resistances of the individual components when connected in series:

$$R_n = R_{an} + R_{bn} + R_{cn} \tag{5}$$

The thermal resistance R_{an} of the surface cladding layer is calculated using the Fourier theorem:

$$R_{an} = \frac{h_a}{\lambda_N \frac{l}{p}} \tag{6}$$

where λ_N is the thermal conductivity of the surface cladding layer; h_a is the surface thickness of the cladding layer; and p is the number of small ideal units.

In order to improve the accuracy of the thermal resistance R_{bn} calculation at the junction, an approximate equivalent thickness is obtained by dividing the length of the upper and bottom edges of the function surface by the definite integral. The contact area is calculated based on the length of the Gaussian distribution function in the region. The thermal resistance R_{bn} at the junction can be obtained as follows:

$$R_{bn} = \frac{\dfrac{\int_{\frac{(n-1)l}{p}}^{\frac{nl}{p}} f(x)dx}{\frac{l}{p}}}{\lambda_N B \cdot \int_{\frac{(n-1)l}{p}}^{\frac{nl}{p}} \sqrt{1 + \left(\frac{df(x)}{dx}\right)^2}\,dx} \tag{7}$$

where B is the width of the whole composite material, and $f(x)$ is Equation (3).

The calculation of substrate thermal resistance R_{cn} is similar to the bonding thermal resistance. The substrate thickness is calculated by subtracting the thickness of the cladding layer and joint from the overall height:

$$R_{cn} = \frac{H - h_a - \dfrac{\int_{\frac{(n-1)l}{p}}^{\frac{nl}{p}} f(x)dx}{\frac{l}{p}}}{\lambda_c B \cdot \int_{-\frac{(n-1)l}{p}}^{\frac{nl}{p}} \sqrt{1 + \left(\frac{df(x)}{dx}\right)^2}\,dx} \tag{8}$$

where λ_c is the thermal conductivity of the substrate, and H is the total thickness of the composite element model.

3.2. Effective Thermal Conductivity Model of Composite Materials

The thermal resistance of the effective element model is obtained using Equations (5)–(8). The physical model of composite materials is composed of multiple element models in parallel, as shown in Figure 6.

We calculate the thermal resistance R_o using the element model, and then calculate the actual thermal conductivity according to the actual-size thermal resistance relationship:

$$R_\alpha = R_o \cdot \frac{2l}{A} \tag{9}$$

R_α is the actual thermal resistance of the whole, and the thermal conductivity λ_α of the composite is calculated according to the Fourier law:

$$\lambda_\alpha = \frac{H}{R_\alpha AB} \tag{10}$$

That is, according to the above formula, the thermal conductivity model Formula (11) of the composite prepared via laser cladding is finally obtained as follows:

$$\lambda_\alpha = \frac{HA \cdot \sum_{n=1}^{p} \left(\dfrac{\lambda_N l}{h_a p} + \dfrac{\lambda_N l B \cdot \int_{\frac{(n-1)l}{p}}^{\frac{nl}{p}} \sqrt{1+\left(\frac{df(x)}{dx}\right)^2}\,dx}{p \cdot \int_{\frac{(n-1)l}{p}}^{\frac{nl}{p}} f(x)dx} + \dfrac{\lambda_c B \cdot \int_{-\frac{(n-1)l}{p}}^{\frac{nl}{p}} \sqrt{1+\left(\frac{df(x)}{dx}\right)^2}\,dx}{H - h_a - \dfrac{p \cdot \int_{\frac{(n-1)l}{p}}^{\frac{nl}{p}} f(x)dx}{l}} \right)}{ABl} \tag{11}$$

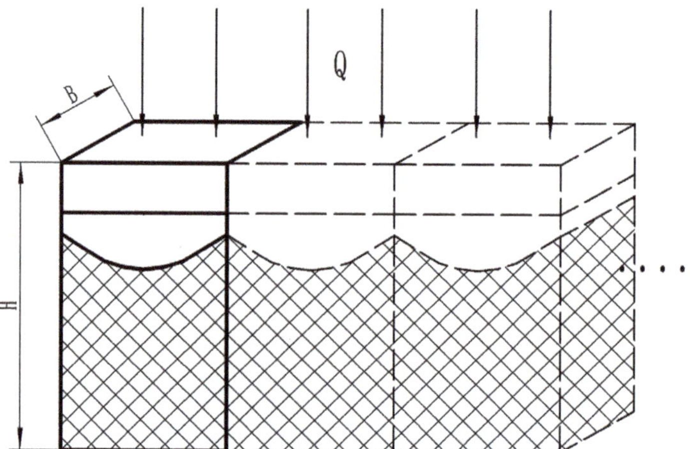

Figure 6. Thermal resistance parallel mode of the effective unit model.

4. Verification and Analysis of Heat Conduction Equation

4.1. Experimental Preparation Method

The accuracy of the model was assessed by comparing numerical calculations with experimental measurements. The test material used was a pure copper sheet metal measuring 100 mm by 60 mm and having a thickness of 6 mm. The laser cladding preparation equipment used the 1064 nm light source produced by Hans laser (Shenzhen, China), the bc104 coaxial powder feeding cladding device produced by RAYtool (Guangdong, China), the powder feeding device produced by Songxing Welding (Guangzhou, China), and the 4-axis CNC machine tool of Siemens control system (Munich, Germany). The process used jp8000 supersonic flame gun produced by Fushan Advanced Surface Technologies Ltd. (Foshan, China) for spraying test.

The surface of the sample was polished with sandpaper and cleaned with anhydrous ethanol. Ni60A powder (75 μm $\sim$ 48 μm) was selected as the cladding material. Notably, the laser's absorption rate on the pure copper surface at this frequency is merely 5%, making it challenging to apply a coating to the copper surface [22,23]. The thermal conductivity of copper reaches 401 W/(m × k), making it challenging to concentrate energy and generate a molten pool. Considering the high laser absorption rate of nickel-based materials, we ensured the safety of personnel and equipment. In this investigation, a layer of Ni60A powder with a thickness of 0.3 mm was initially deposited onto the copper plate through supersonic thermal spraying, as depicted in Figure 7 [24]. The process parameters are shown in Table 1.

Figure 7. Surface and section of pure copper via supersonic thermal spraying.

Table 1. Process parameters of HVOF on copper surface.

Technique	Powder Drying Temperature (°C)	Kerosene Flow Rate (L/h)	Oxygen Flow Rate (L/min)	Nitrogen Flow Rate (L/min)	Powder Feed Rate (L/min)	Spray Distance (mm)	Linear Torch Velocity (mm/s)
HVOF	300	19	750	14	65	310	600

The thickness of the Ni60A preset layer is thin, which is influenced by the high thermal conductivity of copper. For the first layer of laser surface cladding, a high-energy concentration is required to generate a molten pool. In the second layer, coaxial powder feeding is employed for laser cladding. Due to the low thermal conductivity of nickel-based materials and the low reflectivity of the laser, molten pools are easily formed, and the power is gradually reduced. The parameters for laser surface alloying and coaxial powder feeding are shown in Table 2. The material surface after multiple stacking according to the above process is shown in Figure 8.

Table 2. Laser surface alloying and laser cladding process parameters.

Layers	Process Method	Power (kW)	Speed (mm/s)	Laser Beam Radius (mm)	Adjacent Distance (mm)	Shielding Gas Ar Flow (L/min)	Powder Feed Rate (RPM)
1	Laser surface alloying	3000	7	1.5	2	12	0
2	Laser cladding	1500	5	1	1	12	50
3	Laser cladding	1400	5	1	1	12	50
4	Laser cladding	1000	5	1	1	12	50
5	Laser cladding	1000	5	1	1	12	50

Figure 8. Composite prepared using the above parameters.

The prepared composite material was processed into a size of 10 mm × 10 mm × 5 mm using a high-precision grinder and wire cutting, with cladding thicknesses of 0.65 mm, 0.75 mm, 0.85 mm, and 0.95 mm, respectively, as shown in Figure 9.

The composite cross-section image was taken using a macro camera, the image of Equation (12) was drawn using a software, and the two images were overlapped. As shown in Figure 10, the interface between the surface coating and the substrate is Gaussian distribution and highly fitted, which has a strong correlation. The correlation between the unit model is also proven, where

$$f(x) = h_b exp\left(-x^2 / (0.5r)^2\right) \tag{12}$$

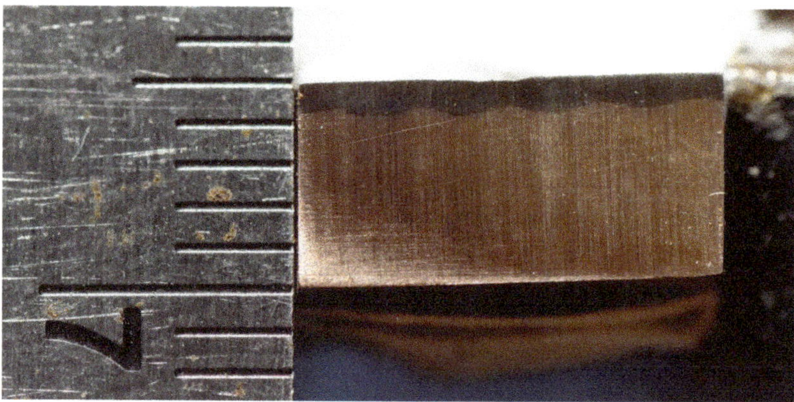

Figure 9. Sample blocks prepared with the above parameters.

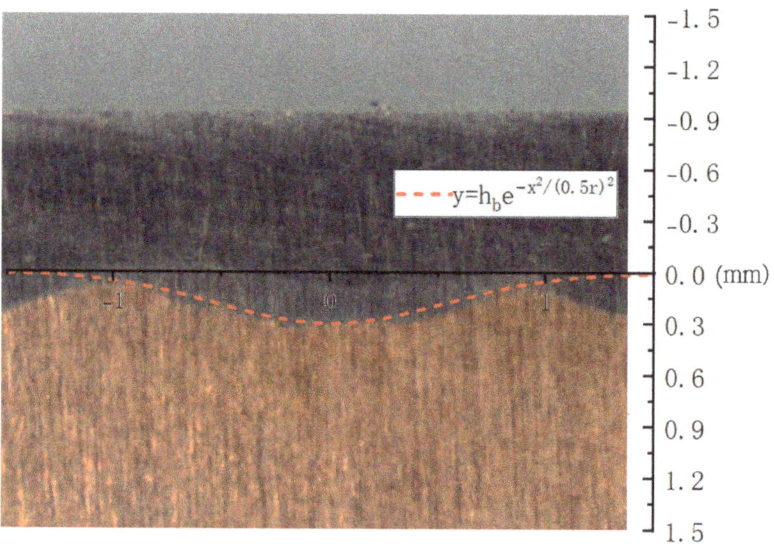

Figure 10. Composite cross section overlapped with Equation (12).

According to the ASTM El461-01 standard [25], the thermal diffusivity of the composite was measured using a laser thermal conductivity meter (LAF427) at different temperatures, and then the sample density was measured using the drainage method. Finally, the thermal conductivity was calculated by using the following formula:

$$\lambda = \alpha \times \rho \times C_\rho \tag{13}$$

where α is the thermal diffusion coefficient; ρ is the density; and C_ρ is the specific heat capacity.

The experimental thermal conductivity was obtained according to the above formula, as shown in Figure 11. According to the results, it can be seen that the effective thermal conductivity of the composite material is 25 W/(m × K) compared with that of the general material die steel H13. Through the establishment of the model, it can be found that the thermal conductivity decreases with an increase in the coating, and the approximate value range of the cladding layer thickness can be determined.

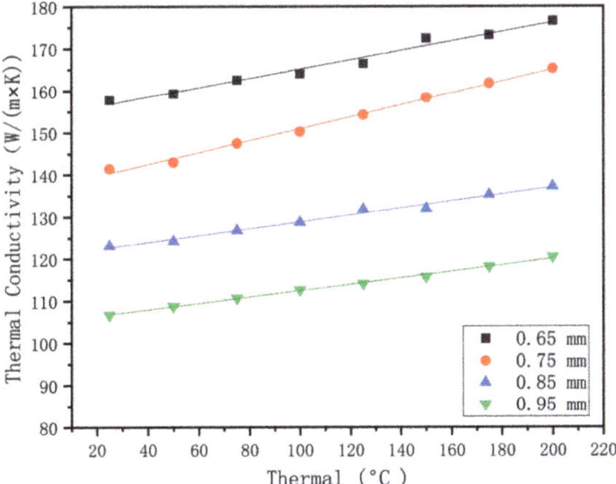

Figure 11. Experimental thermal conductivity of composite at different temperatures.

4.2. Model Comparison and Analysis

According to the preparation process, the following prediction and calculation parameters were selected: the radius of the laser beam is 1.5 mm, the distance between two adjacent laser cladding layers is 2 mm, the height of the first laser cladding layer is 0.3 mm, and the number of small ideal units is 100. It is verified according to the experimental material parameters that the cladding layer uses powder metal containing the content shown in Table 3.

Table 3. Content of Ni60A powder.

	Chemical Composition (wt.%)					
Material name	Ni	C	Si	B	Cr	Fe
Ni60A	Bai	0.9	4	3.2	16	5

The thermal conductivity of Ni60A material was derived using JMatPro 6.0 (Sente Software Ltd., Guildford, UK) software, and the thermal conductivity is shown in Figure 12.

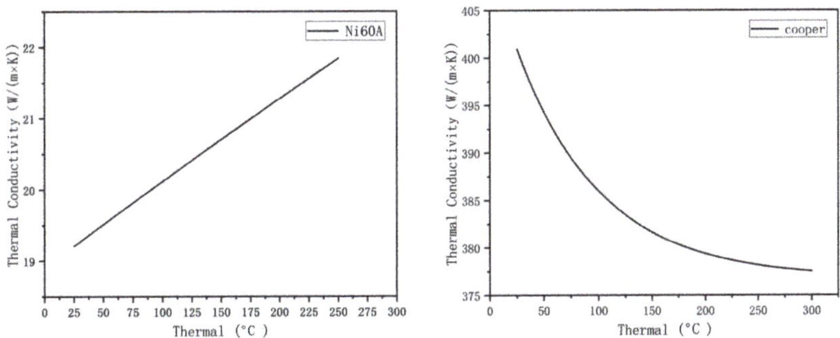

Figure 12. Thermal conductivity of Ni60A and copper at different temperatures.

The model's predicted values were compared with the experimental measured values using Equation (11). As shown in Figure 13, the effective thermal conductivity of different cladding layer thicknesses increases with temperature. The overall average accuracy of the predicted results of the mathematical model is 82.14%, and the overall trend is in good agreement with the experimental data. The accuracy increases with an increase in coating thickness.

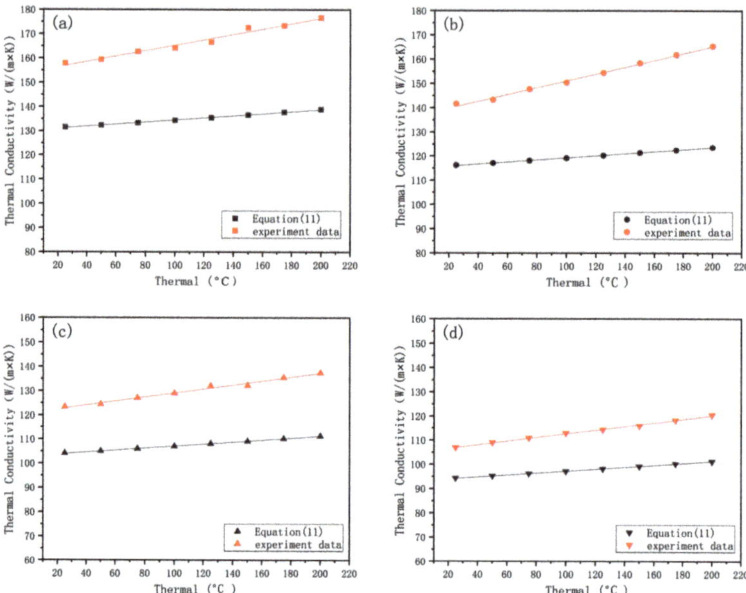

Figure 13. Predicted and measured values of effective thermal conductivity at different temperatures. Effective thermal conductivity of the cladding layer with thicknesses of (**a**) 0.65 mm, (**b**) 0.75 mm, (**c**) 0.85 mm, and (**d**) 0.95 mm.

The model error analysis shows that the overall mathematical model has a lower deviation, and the error increases with a decrease in coating thickness. The main reason for the lower deviation is that the thermal conductivity of the material used in the model is lower than the actual situation. To ensure its universality, the two models ignore the alloy produced by combining two dissimilar alloy materials in the laser cladding process. In reality, the energy-dispersive spectrometer detects the joint of the two materials, as shown in Figure 14. There is a part of metallurgical bonding at the junction. According to the research conducted by Bai et al. [26], the thermal conductivity between the Ni60A powder and the copper substrate increases with an increase in copper content and temperature. When the copper content was 30%, the composite showed a higher thermal conductivity (26.8 W/(m × K)) than the thermal conductivity of Ni60A set in the prediction model. Therefore, the overall prediction value of the mathematical model has lower deviation, and the error increases with a reduction in the thickness of the cladding layer.

To study its influence, the thermal conductivity of Cu/Ni alloy is introduced into the model Formula (11). The thermal conductivity of the composite material is calculated as follows:

$$\lambda_\alpha = \frac{HA \cdot \sum_{n=1}^{p} \left(\frac{\lambda_N l}{h_a p} + \frac{\lambda_{c\_n} l B \cdot \int_{\frac{(n-1)l}{p}}^{\frac{nl}{p}} \sqrt{1+\left(\frac{df(x)}{dx}\right)^2}\,dx}{p \cdot \int_{\frac{(n-1)l}{p}}^{\frac{nl}{p}} f(x)dx} + \frac{\lambda_c B \cdot \int_{\frac{(n-1)l}{p}}^{\frac{nl}{p}} \sqrt{1+\left(\frac{df(x)}{dx}\right)^2}\,dx}{H-h_a-\frac{p \cdot \int_{\frac{(n-1)l}{p}}^{\frac{nl}{p}} f(x)dx}{l}} \right)}{ABl} \tag{14}$$

where λ_{c_n} is the thermal conductivity of Cu/Ni alloy.

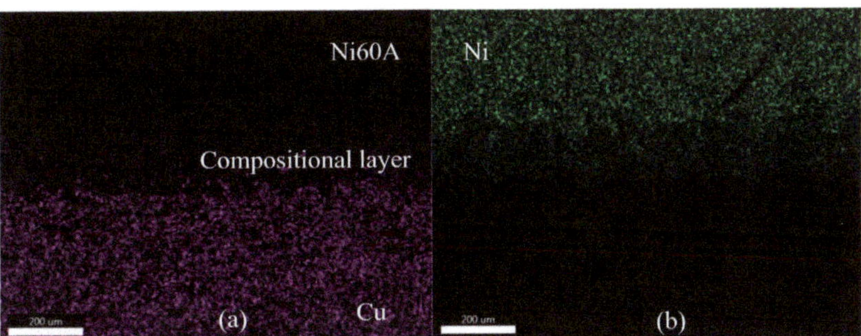

Figure 14. Image obtained when scanning the junction using energy-dispersive spectrometer: (**a**) Cu to (**b**) Ni through the interfaces.

According to the thermal conductivity (26.8 W/(m × K)) in [26], the calculated results of Formula (14) were compared with the data in Figure 13. The comparison of the results is shown in Figure 15. The accuracy of the overall model (Formula (15)) improved by 4.97% after considering metallurgical integration through data comparison, as shown in Figure 15.

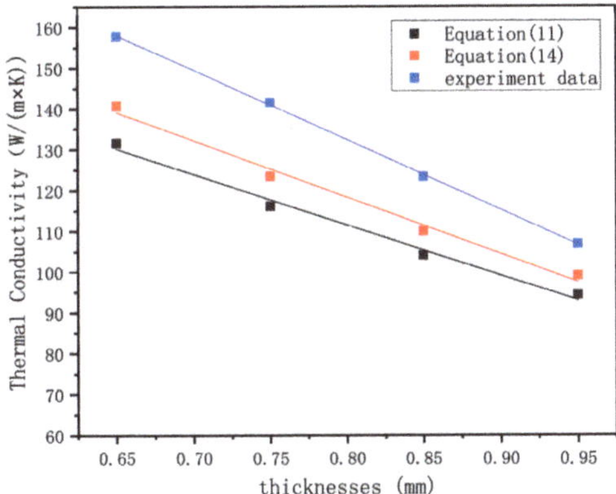

Figure 15. Model comparison after considering the thermal conductivity of the transition layer.

Through the above analysis, it can be found that the systematic error comes from underestimating the thermal conductivity of the sediment. The main impact is in the transition-layer boundary, the size and element distribution of which are affected by multiple factors K. Halmesova et al. [14] found that the use of different process parameters during laser cladding preparation resulted in different metallurgical bonding areas between the two materials, which affected the thermal conductivity. The metallurgical bonding areas under different parameters need to be retested and calculated. Bonny Onuike et al. [2] found that the use of laser cladding to produce materials resulted in a finer grain structure compared to traditional production techniques, resulting in different thermal conductivity coefficients for the same material. In order to improve the accuracy of the model, it is necessary to measure the thermal conductivity of each part of the composite material preparation process to improve its accuracy. This is clearly contrary to the purpose of establishing a model to improve the design speed of composite materials. To ensure that the model has some guiding significance for material design, this study simplifies the transition boundary to a certain extent.

Through a comparison of the model established in this study, it was found that the relationship between thermal conductivity and key parameters varied positively with changes in key parameters. The model can be used to roughly predict thermal conductivity under different parameters. At the same time, the key parameters of composite materials can be designed based on actual needs. Figure 16 shows that the effective thermal conductivity decreases by 75% logarithmically as the thickness of the surface cladding layer increases from 0.5 mm to 2.5 mm. More specifically, when the spot diameter is 3 mm and the overlap rate is 33.3%, the effective thermal conductivity decreases logarithmically, with the thickness increasing from 0.5 mm to 2.5 mm. This factor has a strong correlation, which should be considered first in future research on materials' thermodynamic and mechanical properties.

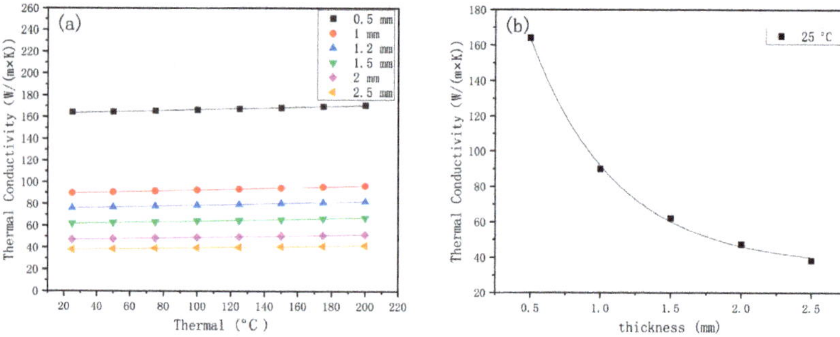

Figure 16. Effective thermal conductivity under different thicknesses. (**a**) Thermal conductivity of composites with different thicknesses (**b**) Thermal conductivity of composites of different thicknesses at 25 °C.

The proposed mathematical model can easily obtain the thermal conductivity of composites prepared using different cladding materials and processes. It provides theoretical and practical guidance for future research on the thermodynamic properties of composites prepared via laser cladding.

5. Conclusions

In this study, a mathematical model was proposed to predict the effective thermal conductivity of laser-clad nickel-based alloy composites on copper surfaces. The model was verified through comparison and experimental data. From the results obtained, we draw the following conclusions:

- By comparing the experimental data, a mathematical model (Equation (11)) was established to predict the effective thermal conductivity, with an accuracy of 82.13%, and the overall trend was in good agreement with the experimental data.
- By comparing the experimental data, the accuracy of the prediction results increases with an increase in the cladding layer thickness, and the overall deviation of the mathematical model results is found. The reason for the overall error is that to ensure its universality, the model ignores the alloy produced by the combination of two different alloy materials in the process of laser cladding. In fact, the metallurgical bonding of Ni/Cu at the joint improves the thermal conductivity of the material. When the thickness of the cladding layer is thin, the effect of metallurgical bonding should be considered.

To sum up, these key parameters not only affect the quality of the coating but also affect the effective thermal conductivity of composites. Therefore, establishing this mathematical model can provide theoretical guidance and optimization direction for the study of materials and the future parameter selection of composite materials.

Author Contributions: Conceptualization, Y.L.; methodology, Y.L.; software, Y.L. and C.L.; validation, Y.L. and C.L.; formal analysis, Y.L.; investigation, C.T.; data curation, C.T.; writing—original draft preparation, Y.L.; writing—review and editing, B.G.M.; supervision, X.C.; project administration, X.C.; funding acquisition, X.C. All authors have read and agreed to the published version of the manuscript.

Funding: This project received funding from key research and industrialization projects of science and technology innovation in Fujian Province, under grant No. 2023XQ005.

Institutional Review Board Statement: Not applicable.

Informed Consent Statement: Not applicable.

Data Availability Statement: The data used to support the findings of this study are included within the article.

Conflicts of Interest: The authors declare no conflict of interest.

References

1. Onuike, B.; Bandyopadhyay, A. Additive manufacturing of Inconel 718—Ti6Al4V bimetallic structures. *Addit. Manuf.* **2018**, *22*, 844–851. [CrossRef]
2. Onuike, B.; Heer, B.; Bandyopadhyay, A. Additive manufacturing of Inconel 718—Copper alloy bimetallic structure using laser engineered net shaping (LENS™). *Addit. Manuf.* **2018**, *21*, 133–140. [CrossRef]
3. Afshari, M.; Hamzekolaei, H.G.; Mohammadi, N.; Yazdanshenas, M.; Hamounpeyma, M.; Afshari, H. Investigating the effect of laser cladding parameters on the microstructure, geometry and temperature changes of Inconel 718 superalloy using the numerical and experimental procedures. *Mater. Today Commun.* **2023**, *35*, 106329. [CrossRef]
4. Marzban, J.; Ghaseminejad, P.; Ahmadzadeh, M.H.; Teimouri, R. Experimental investigation and statistical optimization of laser surface cladding parameters. *Int. J. Adv. Manuf. Technol.* **2015**, *76*, 1163–1172. [CrossRef]
5. Lamikiz, A.; Tabernero, I.; Ukar, E.; de Lacalle, L.; Delgado, J. Influence of the Laser Cladding Strategies on the Mechanical Properties of Inconel 718. In Proceedings of the International Conference on Advances in Materials and Processing Technologies, Paris, France, 24–27 October 2010; Chinesta, F., Chastel, Y., El Mansori, M., Eds.; Volume 1315, pp. 1576–1581. [CrossRef]
6. Pascu, A.; Stanciu, E.M.; Croitoru, C.; Roata, I.C.; Tierean, M.H. Pulsed Laser Cladding of Ni Based Powder. In Proceedings of the International Conference on Innovative Research—Icir Euroinvent 2017, Iasi, Romania, 25–26 May 2017.
7. Vinoth, S.M.; Babu, P.D.; Marimuthu, P.; Phalke, S.S. Laser Cladding of Nickel Powder on AISI 202 Stainless Steel and Optimization of the Process Parameters. In *Advances in Manufacturing Processes, Select Proceedings of ICEMMM 2018*; Sekar, K., Gupta, M., Arockiarajan, A., Eds.; Springer: Berlin/Heidelberg, Germany, 2019; pp. 197–203.
8. Riquelme, A.; Rodrigo, P.; Escalera-Rodríguez, M.D.; Rams, J. Analysis and optimization of process parameters in Al-SiCp laser cladding. *Opt. Laser Eng.* **2016**, *78*, 165–173. [CrossRef]
9. Garcia-Herrera, J.E.; Henao, J.; Espinosa-Arbelaez, D.G.; Gonzalez-Carmona, J.M.; Felix-Martinez, C.; Santos-Fernandez, R.; Corona-Castuera, J.; Poblano-Salas, C.A.; Alvarado-Orozco, J.M. Laser Cladding Deposition of a Fe-based Metallic Glass on 304 Stainless Steel Substrates. *J. Therm. Spray Technol.* **2022**, *31*, 968–979. [CrossRef]
10. Siddiqui, A.A.; Dubey, A.K. Recent trends in laser cladding and surface alloying. *Opt. Laser Technol.* **2021**, *134*, 106619. [CrossRef]
11. Cao, S.; Liang, J.; Wang, L.; Zhou, J. Effects of NiCr intermediate layer on microstructure and tribological property of laser cladding Cr3C2 reinforced Ni60A-Ag composite coating on copper alloy. *Opt. Laser Technol.* **2021**, *142*, 106963. [CrossRef]
12. Bochenek, K.; Węglewski, W.; Strojny-Nędza, A.; Pietrzak, K.; Chmielewski, T.; Chmielewski, M.; Basista, M. Microstructure, Mechanical, and Wear Properties of NiCr-Re-Al₂O₃ Coatings Deposited by HVOF, Atmospheric Plasma Spraying, and Laser Cladding. *J. Therm. Spray Technol.* **2022**, *31*, 1609–1633. [CrossRef]
13. Chen, Q.; Yu, M.; Cao, K.; Chen, H. Thermal conductivity and wear resistance of cold sprayed Cu-ceramic phase composite coating. *Surf. Coat. Technol.* **2022**, *434*, 128135. [CrossRef]
14. Halmešová, K.; Trojanová, Z.; Koukolíková, M.; Brázda, M.; Džugan, J.; Huang, W.C. Effect of laser power on thermal properties of multimaterial structure Inconel 718 and stainless steel 316L processed by directed energy deposition. *J. Alloys Compd.* **2022**, *927*, 167082. [CrossRef]
15. Yang, M.; Zhu, Y.; Wang, X.; Wang, Q.; Ai, L.; Zhao, L.; Chu, Y. A novel low thermal conductivity thermal barrier coating at super high temperature. *Appl. Surf. Sci.* **2019**, *497*, 143774. [CrossRef]
16. He, F.; Wang, Y.; Zheng, W.; Wu, J.; Huang, Y. Effective thermal conductivity model of aerogel thermal insulation composite. *Int. J. Therm. Sci.* **2022**, *179*, 107654. [CrossRef]
17. Yang, Y.; Zhang, T.; Reddy, K.R.; Li, J.; Liu, S. Thermal conductivity of scrap tire rubber-sand composite as insulating material: Experimental investigation and predictive modeling. *Constr. Build. Mater.* **2022**, *332*, 127387. [CrossRef]
18. Yan, B.; Cheng, L.; Li, B.; Liu, P.; Wang, X.; Gao, R.; Yang, Z.; Xu, S.; Ding, X.; Zhang, P. Bi-directional prediction of structural characteristics and effective thermal conductivities of composite fuels through learning from finite element simulation results. *Mater. Des.* **2020**, *189*, 108483. [CrossRef]
19. Węglewski, W.; Pitchai, P.; Chmielewski, M.; Guruprasad, P.J.; Basista, M. Thermal conductivity of Cu-matrix composites reinforced with coated SiC particles: Numerical modeling and experimental verification. *Int. J. Heat Mass Transf.* **2022**, *188*, 122633. [CrossRef]
20. Li, B.; Du, J.; Sun, Y.; Zhang, S.; Zhang, Q. On the importance of heat source model determination for numerical modeling of selective laser melting of IN625. *Opt. Laser Technol.* **2023**, *158*, 108806. [CrossRef]

21. Dai, D.; Gu, D. Thermal behavior and densification mechanism during selective laser melting of copper matrix composites: Simulation and experiments. *Mater. Des.* **2014**, *55*, 482–491. [CrossRef]
22. Gao, M.; Li, S.; Guan, W.; Xie, H.; Wang, X.; Liu, J.; Wang, H. Excellent thermal shock resistance of NiCrAlY coatings on copper substrate via laser cladding. *J. Mater. Sci. Technol.* **2022**, *130*, 93–102. [CrossRef]
23. Qi, Z.; Chen, C.; Wang, C.; Zhou, Z.; Zhou, J.; Long, Y. Effect of different laser wavelengths on laser cladding of pure copper. *Surf. Coat. Technol.* **2023**, *454*, 129181. [CrossRef]
24. Imran, M.K.; Masood, S.H.; Brandt, M.; Bhattacharya, S.; Mazumder, J. Direct metal deposition (DMD) of H13 tool steel on copper alloy substrate: Evaluation of mechanical properties. *Mater. Sci. Eng. A* **2011**, *528*, 3342–3349. [CrossRef]
25. *ASTM G146-01*; Standard Practice for Evaluation of Disbonding of Bimetallic Stainless Alloy/Steel Plate for Use in High-Pressure, High-Temperature Refinery Hydrogen Service. ASTM: West Conshehoken, PA, USA, 2018.
26. Bai, D.; Zhang, C.; Chen, Z.; Zhang, Y.; Li, G.; Lu, X. Thermal conductivity characterization of Ni60A/WC composites with different copper dilutions. *Mater. Lett.* **2022**, *311*, 131546. [CrossRef]

materials

Article

Maintaining Excellent Mechanical Properties via Additive Manufacturing of Low-N 25Cr-Type Duplex Stainless Steel

Jianguo He, Jiesheng Lv, Zhigang Song *, Changjun Wang, Han Feng, Xiaohan Wu, Yuliang Zhu and Wenjie Zheng

Research Institute of Special Steels, Central Iron & Steel Research Institute Co., Ltd., Beijing 100081, China; hejianguo@nercast.com (J.H.); lvllvlv@foxmail.com (J.L.); wangchangjun@nercast.com (C.W.); fenghan@nercast.com (H.F.); wuxiaohan@nercast.com (X.W.); zhuyuliang@nercast.com (Y.Z.); zhengwenjie@nercast.com (W.Z.)
* Correspondence: songzhigang@nercast.com

Abstract: Duplex stainless steel (DSS) exhibits good mechanical properties and corrosion resistance, and has attracted more and more attention within the fields of both science and technology. However, the increasing levels of N and of Cr, Mo, etc., as alloying elements in DSS increase production difficulty. In particular, the N element increases the risk of Cr_2N precipitation, which can seriously deteriorate the thermal plasticity of DSS, while increasing its strength. For this reason, a low-N-content 25Cr-type DSS was designed in order to adapt additive manufacturing processes. With regard to the nano-inclusions of oxide precipitation and effective grain refinement, and considering the benefits of selective laser melting fabrication, a low-N 25Cr-type duplex stainless steel with a 0.09 wt.% N content achieved high mechanical properties, with a yield strength of 712 MPa and an elongation of 27.5%, while the V-notch impact toughness was 160 J/cm^2. The microstructure evolution and the reasons behind the improvement in mechanical properties will be discussed in detail.

Keywords: selective laser melting; duplex stainless steels; nano-inclusion; microstructure; mechanical properties

check for **updates**

Citation: He, J.; Lv, J.; Song, Z.; Wang, C.; Feng, H.; Wu, X.; Zhu, Y.; Zheng, W. Maintaining Excellent Mechanical Properties via Additive Manufacturing of Low-N 25Cr-Type Duplex Stainless Steel. *Materials* **2023**, *16*, 7125. https://doi.org/10.3390/ma16227125

Academic Editor: Francesco Iacoviello

Received: 29 August 2023
Revised: 18 September 2023
Accepted: 21 September 2023
Published: 10 November 2023

1. Introduction

Duplex stainless steel (DSS) is composed of ferrite and austenite phases that exhibit a higher strength than conventional austenitic stainless steel, while the higher Cr, Mo, and N concentrations in DSS contribute to an increased and outstanding corrosion resistance compared to that of conventional austenitic stainless steel [1–7]. DSS also exhibits a unique resistance to intergranular corrosion in certain environments, making it irreplaceable in the petroleum, chemical, and marine [8–11] industries, among others. Compared to the 300-series austenitic stainless steels, duplex stainless steel can achieve higher strength and PREN values at similar or lower costs.

Although duplex stainless steel has certain advantages over austenitic stainless steel in terms of performance, there are still challenges facing the complete replacement of austenitic stainless steel by duplex stainless steel on the market when it comes to widely used application. The American Society for Testing Materials (ASTM) only includes seven types of cast duplex stainless steels in its standards, while there are hundreds of types of forged duplex stainless steels [8]. The slow cooling process during the cooling process after solidification leads to the formation of harmful secondary precipitates, such as CrN, Cr_2N, sigma (σ) phase, or chi(χ) phase, which significantly affect both the mechanical properties and the corrosion resistance [11–13]. This makes it challenging to maintain the performance of duplex stainless steel in complex structural applications using traditional casting processes, which is an important reason behind the limited scope of application for duplex stainless steel.

In recent years, the development of additive manufacturing technology has introduced a new metal-forming process with extremely fast cooling rates [14–21]. The one-time forming method reduces spending on molds and allows for the manufacturing of more complex structural components, while avoiding issues like σ phase precipitation and grain coarsening during casting cooling [19,22–28]. The characteristics of additive manufacturing effectively address the issues of casting formation and the diversification of product forms in duplex stainless steel.

Based on the aforementioned research background, this work focuses on 2507-type duplex stainless steel with a lower N content (≤0.1 wt.%) and utilizes the laser powder bed fusion (L-PBF) method to fabricating 25Cr-type DSS with a high strength and plasticity and excellent toughness. The evolution of its microstructure at different stages is analyzed, and precipitation-strengthening and grain-refinement advantages are achieved through control of the composition and printing methods. This method is expected to provide an alternative composition optimization strategy for additive manufacturing processes for the application of duplex stainless steel.

2. Materials and Methods

Argon-gas atomization was used to prepare 25Cr-type DSS powder with the size distribution of 15–50 μm. The main chemical composition of the powder was measured via inductively coupled plasma atomic emission spectrometry (ICP-AES), and the results are shown in Table 1. Figure 1a shows the morphology observed under scanning electron microscopy (SEM); the powder exhibits a smooth and round shape without satellite particles. The X-ray diffraction (XRD) analysis results for the powder are shown in Figure 1b, revealing a phase composition of 99% ferrite and 1% austenite, with no harmful phases detected.

Table 1. The chemical composition of the 25Cr-type DSS powder [1].

Cr	Ni	Mo	N	Mn	O	Si	C	P	S
24.70	6.52	3.74	0.098	0.55	0.028	0.35	0.0050	0.0058	0.0029

[1] S, C, and O elements were determined via the infrared absorption method.

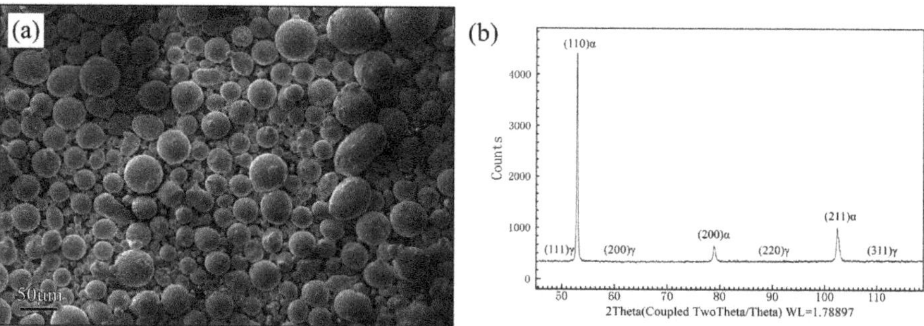

Figure 1. Powder morphology and phase information: (**a**) powder SEM image, (**b**) XRD diffraction pattern.

The aforementioned powder was used for selective laser melting (SLM) on a DLM-280 metal 3D printer. The build process was performed on a 316 stainless steel substrate in a protective atmosphere of high-purity argon gas (99.9%). The specific sintering parameters were as follows: a laser input power (P) of 190 W, a laser spot diameter of 0.1 mm, a powder layer thickness (h) of 0.02 mm, line spacing (t) of 0.1 mm, a scanning speed (v) of 850 mm/s, and a bidirectional scanning pattern with a 45° angle between each layer. The calculated energy density, using Formula (1), was 117.65 J/cm³.

$$E = \frac{P}{v \cdot h \cdot t} \tag{1}$$

The built specimens were subjected to solid solution treatment at temperatures of 1100, 1150, and 1200 °C for 1 h each. Cube samples measuring $10 \times 10 \times 5$ mm were used to characterize the microstructure, while rough impact samples with dimensions of $10 \times 60 \times 6$ mm and rough tensile samples with dimensions of $20 \times 70 \times 6$ mm were used to test the mechanical properties. The final dimensions of the impact and tensile specimens are shown in Figure 2a,b, respectively. The relative density of the specimens, using the Archimedes drainage method, was 99.87%.

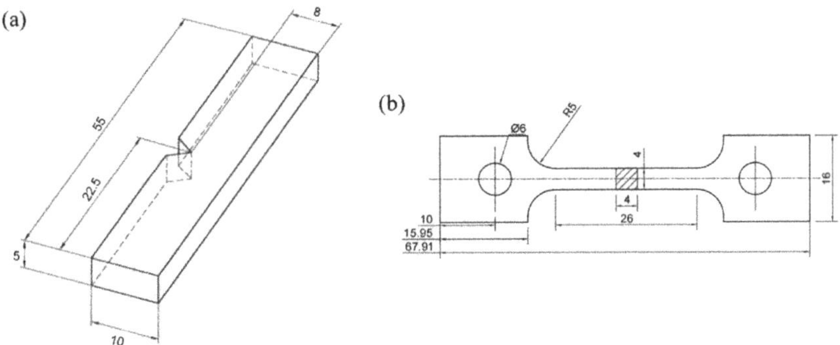

Figure 2. The dimensions of the specimens for testing mechanical properties: (**a**) the impact specimen dimensions; (**b**) the tensile specimen dimensions (unit: mm).

Tensile testing at room temperature was performed using a WE300B tensile testing machine (supplied by Marxtest Technology Co., Ltd., Jinan, China) with a strain rate of 1×10^{-3} s^{-1}. Impact testing at room temperature was conducted using a NI750 metal pendulum impact testing machine (supplied by NCS Testing Technology Co., Ltd., Beijing, China). After mechanical grinding and polishing, the square samples were immersed in a potassium permanganate solution at 50 °C for 3 h, and their microstructure was observed using a LEICA MEF4M optical microscope (supplied by Leica Microsystems Shanghai Ltd., Shanghai, China). The backscattered electron diffraction (EBSD) samples were immersed in a 10% alcoholic hydrochloric acid solution and subjected to electrolytic polishing at a voltage of 25 V for 30 s. EBSD characterization was carried out using a FEI Quanta650 field-emission scanning electron microscope (supplied by FEI Co., Ltd., Hillsboro, OR, USA), and the data were processed using Channel 5 software. Transmission electron microscope (TEM) samples were mechanically thinned to a thickness of 40 μm using silicon carbide paper and then further thinned using a dual-jet electropolisher at a voltage of -28 V and a temperature of -20 °C. The electrolyte consisted of 10% hydrochloric acid and 90% anhydrous ethanol. Observation was performed using a FEI TECNAI G2 F20 transmission electron microscope (supplied by FEI Co., Ltd., Hillsboro, American) operating at an accelerating voltage of 200 kV. X-ray diffraction experiments on the samples were conducted using a Bruker D8 Advance X-ray diffractometer (supplied by Bruker Co., Ltd., Billerica, MA, USA) with a tube voltage of 35 kV, a tube current of 40 mA, an incident wavelength of $\lambda = 0.179$ nm, and a scanning speed of 2°/min. The nitrogen content in the printed specimens was measured using an NCS ONH-5500 analyzer (supplied by NCS Testing Technology Co., Ltd., Beijing, China).

3. Results and Discussion

3.1. Microstructure

Figure 3a shows the microstructure of the built cubic specimen without heat treatment, which reveals a mosaic-like regular structure. The magnified metallographic image in Figure 3b further reveals that the mosaic-like grid structure is arranged in a regular pattern, with 100 μm as the minimum structural unit. The core of the mosaic unit consists of larger grains, while the edges consist of smaller grains similar to recrystallized grains. The track width formed by the units is 100 μm, arranged in two directions at a 90° angle to each other. The aforementioned mosaic structure is mainly due to the special scanning method, where the laser spot radius is 100 μm, and the scan line spacing is also 100 μm. Due to the higher energy at the core of the laser, the powder in the core receives more energy compared to the powder at the edges, resulting in a difference in grain size between the core and the edges. At this point, the grain size distribution presents larger grains in the core and smaller grains on the sides. As the scanning progresses, after the completion of one layer, the scanning path of the next layer of powder is perpendicular to the previous layer at 90°. The laser energy input of the next layer, to a certain extent, affects the already-formed matrix in the previous layer, causing secondary heating and recrystallization, ultimately leading to the grid-like structure. The microstructure metallography of the specimen when heat-treated at 1200 °C for 1 h is shown in Figure 3c, and it is worth noting that the grid-like mosaic structure in the microstructure remains intact and regular after heat treatment, which is significantly different from the destruction of the grid-like structure that has been observed in previous studies after heat treatment [29–33]. The magnified microstructure shown in Figure 3d reveals that a certain amount of new phase has precipitated on the grain boundaries of the matrix phase and, even after 1 h of heat treatment at 1200 °C, the size remains small. The above microstructural characteristics are mainly attributed to the unique low-nitrogen (N)-composition design.

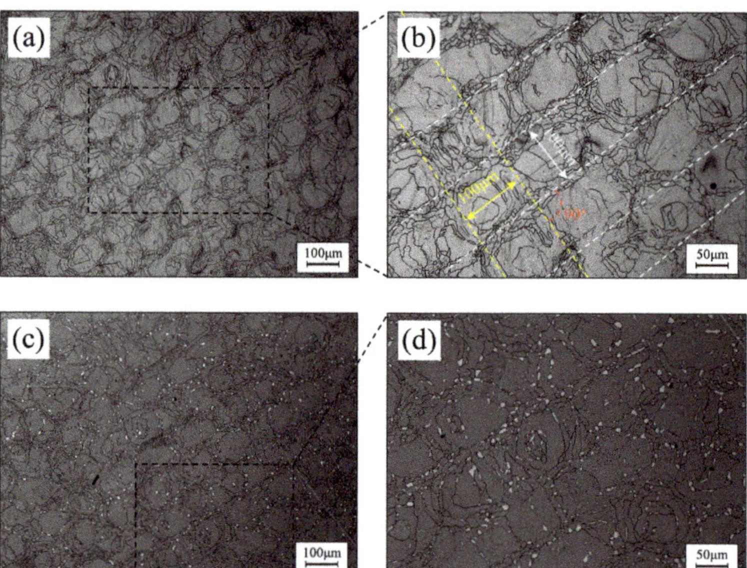

Figure 3. An optical microscope (OM) image of the specimen before and after heat treatment: (**a**) the specimen before heat treatment at 100× magnification, (**b**) the magnified microstructure of the selected area from (**a**), (**c**) the specimen heat-treated at 1200 °C for 1 h at 100× magnification, (**d**) the magnified microstructure of the selected area from (**c**).

To further characterize the microstructural features before and after heat treatment, the phase distribution before and after heat treatment was statistically analyzed using EBSD. Figure 4a shows the EBSD phase map of the laser-irradiated surface before heat treatment, which predominantly consists of the ferrite phase. This is mainly due to the high temperature during the scanning process and the extremely fast cooling rate, which does not allow for the precipitation of austenite during the forming and cooling process. Figure 4b shows the EBSD phase map of the non-heat-treated scanning side, and it can be observed that each layer of ferrite has been melted by the laser into a solid structure with a width of approximately 100 μm. This solid structure will improve the impact toughness, which will be discussed in detail later. Figure 4c,e, respectively, show the phase distribution maps of the scanned front and side after 1 h of heat treatment at 1200 °C. A large amount of austenite precipitates at the grain boundaries of the ferrite, and some austenite precipitates at the low-angle grain boundaries within the ferrite. It can be observed that the size of the austenite precipitated on the high-angle grain boundaries is significantly larger than that precipitated on the low-angle grain boundaries within the grain. This is mainly because the high-angle grain boundaries have a higher energy, providing convenience for the rapid nucleation and growth of austenite. Figure 4d shows the phase distribution map of the scanned front after heat treatment at a lower solution temperature of 1100 °C for 1 h. It can be found that, compared to at 1200 °C, the amount of austenite increases significantly, and the austenite is distributed in a strip-like pattern connected to the grain boundaries of the ferrite. At a lower solution temperature, the morphological characteristics of the austenite phase have changed significantly compared to the discontinuous island-like distribution at a higher solution temperature.

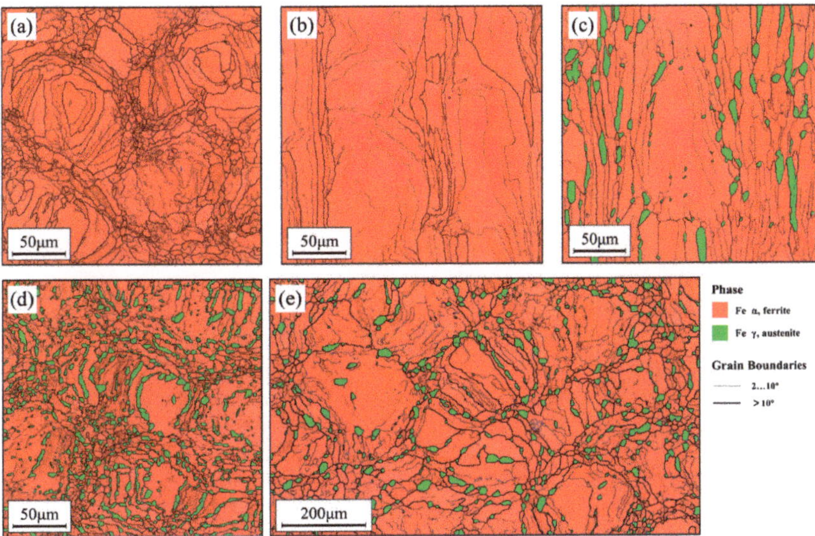

Figure 4. EBSD phase distribution maps: the front surface (**a**) and side surface (**b**) of the specimen before heat treatment, the front surface (**e**) and side surface (**c**) of the specimen heat-treated at 1200 °C for 1 h, the scan of the front surface (**d**) of the specimen heat-treated at 1100 °C for 1 h.

3.2. Mechanical Properties

3.2.1. Tensile Properties

Figure 5 shows the tensile properties of the sample before and after heat treatment. It can be observed that the sample has a very high yield strength (920 MPa) and tensile strength (922 MPa) in the state before heat treatment, but the elongation is only 2%. Concerning the tensile properties of the sample after heat treatment, it can be observed that, as the solution temperature increases, the strength and plasticity of the sample simultaneously increase. After solution treatment at 1200 °C, even with only a weak solid solution strengthening due to a N content of only 0.098%, the yield strength is still as high as 712 MPa. In previous studies, the yield strength of standard-grade (higher-N-content, compared to this paper) 2205 and 2507 duplex stainless steel, whether through casting or additive manufacturing, typically ranged from 600 to 660 MPa [15,34–36]. The test results obtained in this paper showed that the yield strength of the samples after a 1 h solution treatment at 1200 °C was significantly higher than that shown in the experimental results of previous studies.

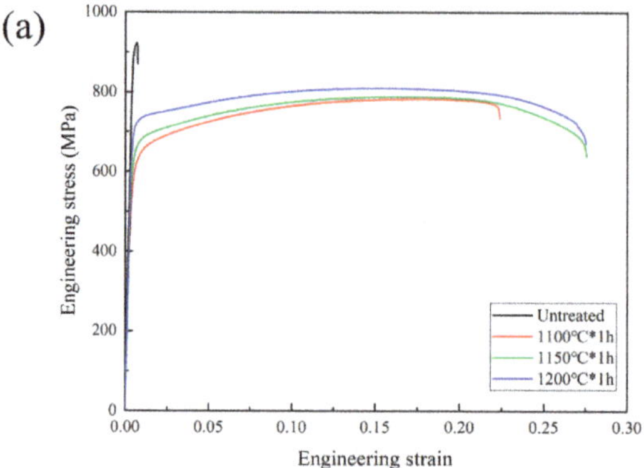

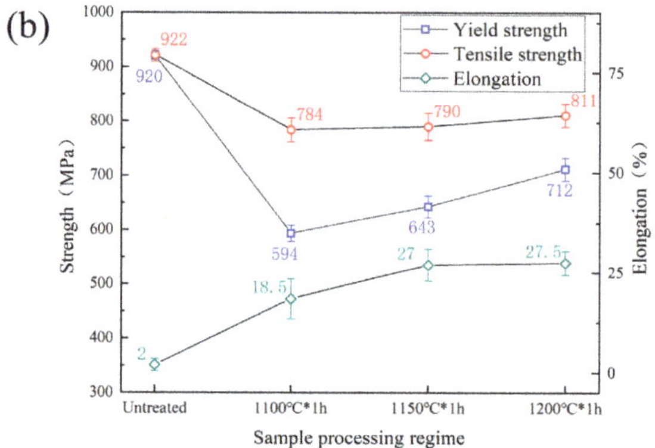

Figure 5. The tensile performance of the specimen: (**a**) the stress–strain curve, (**b**) the yield strength, tensile strength, and elongation at different treatment states.

3.2.2. Impact Properties

Table 2 presents the impact toughness values of the samples before and after heat treatment, compared with the impact performance of the 2707 BPF samples with similar processes. It can be observed that the impact toughness of the experimental steel in this study is significantly higher than that of the reference, and that the impact toughness value of the untreated samples is significantly higher than that in the results for the reference. Although the samples without heat treatment only show a 2% elongation during the tensile process, they exhibit high toughness values. SEM images of the fractured surfaces of the untreated samples at different scales are shown in Figure 6a–c. The unique fracture morphology is an important reason behind the higher toughness value of the samples. From the macroscopic fracture morphology in Figure 6a, it can be observed that the fracture surface of the sample is divided into two sides. The side near the V-notch (the whiter side) shows irregular cleavage fracture, while the other side exhibits a mosaic-like structure similar to its microstructure. The magnified images in Figure 6b,c reveal more clearly the complex fracture forms. In each small structural unit, the central large grains exhibit smooth cleavage fracture planes. The small grain area at the edge of the structural unit consists of a certain number of smaller dimples. The smaller grain size and multiple interfaces in this region partly hinder the propagation of cracks during the fracturing process, thus improving the impact resistance. On the other hand, as mentioned earlier, the microstructure on the sample's side consists of ferrite arranged vertically as a whole. Cracks can only propagate through transgranular fracture within these ferrite regions, resulting in a fracture surface that is similar in appearance to the front surface's grid-like structure. The coexistence of mixed microstructures with different grain sizes, and the integrity of the ferrite in the side structure, contribute to the complex fracture form and the high toughness value of the samples.

Table 2. V-notch impact toughness.

	Impact Toughness Value (aKV), J/cm^2			
	Untreated	1100 °C × 1 h	1150 °C × 1 h	1200 °C × 1 h
This paper	98.75	158.75	161.25	160
Previous studies [26]	18	-	132	-

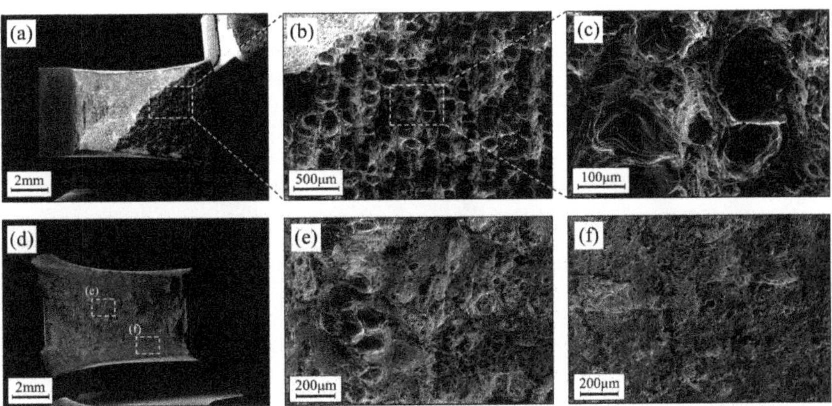

Figure 6. SEM images of the impact fracture surfaces: the fracture morphology of the specimen before heat treatment (**a**) and selected-area enlargement (**b**,**c**), the fracture morphology of the specimen heat-treated at 1200 ° for 1 h (**d**), and selected-area enlargement (**e**,**f**).

The SEM images of the fractured surface of the heat-treated impact samples are shown in Figure 6d–f. The macroscopic image in Figure 6d reveals a composite structure with smooth edges and an uneven core. The magnified image in Figure 6e of the core region shows a combination of dimples and cleavage planes, with some areas still exhibiting grid-like structures. The magnified image of the fracture edge in Figure 6f reveals cleavage planes. The structural differences between different regions of the fracture surface indicate a relatively poor continuity of crack propagation during the fracturing process, ultimately resulting in a more objective measure of the impact toughness.

4. Discussion

The extreme tensile behavior of the untreated sample is mainly due to the presence of a ferrite structure in the untreated state. Typically, at room temperature, the nitrogen saturation solubility in ferrite is 0.007%, while most of the 0.098% nitrogen in the composition is dissolved in the ferrite, leading to a distorted ferrite lattice due to nitrogen oversaturation. Additionally, the repeated heating and rapid cooling during the printing process result in significant residual thermal stresses. The combination of these factors leads to a low plasticity and high strength in the untreated sample. The TEM characterization of the untreated sample confirms the above inference. Figure 7a shows the internal structure of the ferrite at the TEM scale. It can be observed that the interior of the undeformed ferrite is highly disordered. The magnified view in Figure 7b clearly shows a large number of randomly distributed and tangled dislocations within the ferrite grain. Even without deformation, the ferrite in this state is under a significant internal stress and, thus, fractures with only a small amount of deformation (2%).

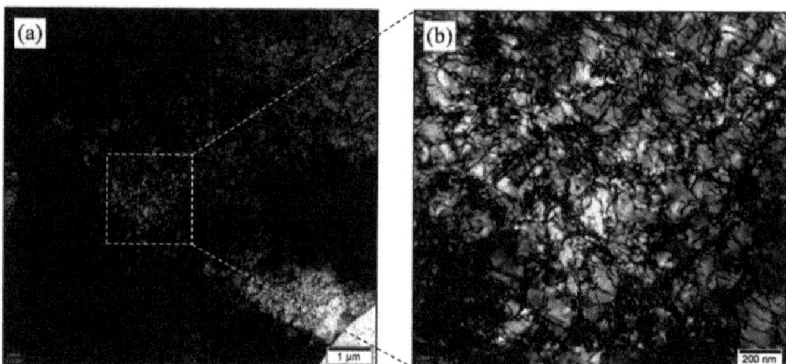

Figure 7. TEM image of the specimen before heat treatment (**a**) and selected-area enlargement (**b**).

As mentioned earlier, the yield strength of the samples after a 1 h heat treatment at 1200 °C reached values as high as 712 MPa, while still maintaining a 27.5% elongation. This anomalous mechanical performance is mainly due to the combined effect of the unique phase morphology of austenite under special composition and the precipitation of high-temperature oxide phases. At lower solution temperatures, the austenite content is higher, and a large amount of austenite forms continuous bands along the ferrite grain boundaries [37,38]. The softer and higher-content austenite leads to a decrease in yield strength, and the large distribution on the grain boundaries easily causes a deformation mismatch between the two phases, leading to fracture. Conversely, at higher solution temperatures, the austenite content is lower, and it is discretely distributed in island-like shapes along the ferrite grain boundaries. For this reason, there is less austenite with a smaller size and discrete distribution, which acts as a second-phase reinforcement by pinning on the ferrite grain boundaries, achieving an enhanced strength and plasticity.

In addition, in previous studies, the introduction of high-temperature oxide Al_2O_3 into the system also leads to an improvement in strength [33,39]. In traditional casting processes, the high-temperature-formed Al_2O_3 easily grows and produces inclusions during slow cooling, making it difficult to eliminate. However, in additive manufacturing processes, the extremely fast cooling rate prevents these high-temperature-formed Al_2O_3 from growing, and they ultimately become precipitates that are beneficial to mechanical properties. In this study, TEM observation also revealed fine dispersed circular Al_2O_3 precipitates, as shown in Figure 8a. The dark-field TEM image in Figure 8b provides a clearer view of the widely varying distribution sizes of Al_2O_3. To better illustrate the types of precipitates, Figure 8c shows the EDS mapping results of the region. It is evident that there is a clustering of Al and O elements in the precipitates. This type of precipitate strengthening is also one of the reasons behind the higher yield strength of the sample. We also characterized the uniform deformation zone of the samples after tensile fracture, using TEM. Figure 8d shows a TEM image of the deformed state, where a large number of dislocations are pinned by small Al_2O_3 inclusions in the ferrite. The interaction between the fine dispersoids of Al_2O_3 particles and dislocations hinders the initial movement of the ferrite during deformation, which contributes to the higher yield strength, compared to that under normal conditions.

In summary, through a unique composition design that breaks away from the conventional 0.24–0.32% nitrogen (N) content in 2507, a unique austenite phase morphology has been induced, while a regular lattice structure is maintained. With the reinforcement of oxide precipitates, the sample exhibits a high yield strength of 712 MPa, while still maintaining a good elongation of 27.5%.

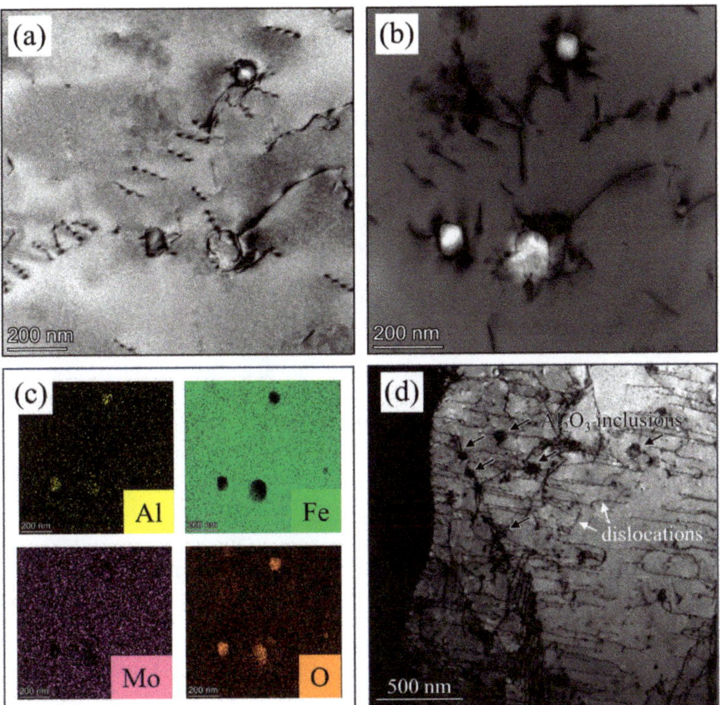

Figure 8. TEM micrographs of oxide inclusions in the specimen after 1 h heat treatment at 1200 °C: (**a**) the undeformed state, (**b**) the dark-field image of (**a,c**) the EDS mapping of (**a,d**) the deformed state.

5. Conclusions

In this study, a low-N 25Cr-type duplex stainless steel for additive manufacturing processes was fabricated through the L-PBF method. The product exhibits an excellent strength and ductility. The main conclusions are as follows:

1. When the strategy of alternating the layer scanning direction of 45° is used, the microstructure of the specimens shows a regular mosaic structure. The microstructure before heat treatment consists of a single ferrite phase. After heat treatment at 1200 °C, discrete and fine austenite precipitates at the ferrite grain boundaries, and the regular mosaic structure remains.

2. The unique morphology of the austenite phase, caused by the special low-N-composition design, along with the strengthening effect of oxide precipitation, leads to a high yield strength of 712 MPa after heat treatment in the experimental steel in this study, while a good elongation of 27.5% is maintained.

3. The specimen of additive-manufacturing low-N 25Cr-type DSS demonstrates an excellent V-notch impact performance, with a toughness value of 98.75 J/cm^2 before heat treatment and approximately 160 J/cm^2 after heat treatment.

Author Contributions: Conceptualization, J.H. and Z.S.; methodology, H.F. and C.W.; validation, Y.Z., W.Z. and X.W.; writing—original draft preparation, J.L.; writing—review and editing, J.H.; supervision, Z.S.; project administration, J.H. All authors have read and agreed to the published version of the manuscript.

Funding: This research was funded by the Innovation Fund of China Steel Research Technology Group Co., Ltd., (No. KNJT05-JT0M-21001).

Institutional Review Board Statement: Not applicable.

Informed Consent Statement: Not applicable.

Data Availability Statement: The data presented in this study are available upon request from the corresponding author.

Conflicts of Interest: The authors declare no conflict of interest.

References

1. Gu, S.; Liu, C.; Kimura, Y.; Yoon, S.; Cui, Y.; Yan, X.; Ju, Y.; Toku, Y. Realizing strength-ductility synergy in a lean duplex stainless steel through enhanced TRIP effect via pulsed electric current treatment. *Mater. Sci. Eng. A* **2023**, *883*, 145534. [CrossRef]
2. Koga, N.; Noguchi, M.; Watanabe, C. Low-temperature tensile properties, deformation and fracture behaviors in the ferrite and austenite duplex stainless steel with various grain sizes. *Mater. Sci. Eng. A* **2023**, *880*, 145354. [CrossRef]
3. Oñate, A.; Toledo, E.; Ramirez, J.; Alvarado, M.I.; Jaramillo, A.; Sanhueza, J.P.; Medina, C.; Melendrez, M.F.; Rojas, D. Production of Nb-doped super duplex stainless steel based on recycled material: A study of the microstructural characterization, corrosion, and mechanical behavior. *Mater. Chem. Phys.* **2023**, *308*, 128294. [CrossRef]
4. Zhang, J.; Zhu, H.; Xi, X.; Li, X.; Xia, Z. Anisotropic response in mechanical and corrosion performances of UNS S31803 duplex stainless steel fabricated by laser powder bed fusion. *J. Mater. Res. Technol.* **2023**, *26*, 4860–4870. [CrossRef]
5. Zhang, X.; Li, J.; Gong, Q.; Liu, C.; Li, J. Deformation behaviors at cryogenic temperature of lean duplex stainless steel with heterogeneous structure prepared by a short process. *J. Mater. Res. Technol.* **2023**, *24*, 9273–9291. [CrossRef]
6. Zhou, F.; Chhun, N.; Cai, Y. Numerical investigation and design of cold-formed lean duplex stainless steel Z-sections undergoing web crippling. *Thin-Walled Struct.* **2023**, *183*, 110324. [CrossRef]
7. Örnek, C.; Mansoor, M.; Larsson, A.; Zhang, F.; Harlow, G.S.; Kroll, R.; Carlà, F.; Hussain, H.; Derin, B.; Kivisäkk, U.; et al. The causation of hydrogen embrittlement of duplex stainless steel: Phase instability of the austenite phase and ductile-to-brittle transition of the ferrite phase—Synergy between experiments and modelling. *Corros. Sci.* **2023**, *217*, 111140. [CrossRef]
8. Francis, R.; Byrne, G. Duplex Stainless Steels—Alloys for the 21st Century. *Metals* **2021**, *11*, 836. [CrossRef]
9. He, L.; Wang, Y.; Singh, P.M. Role of Ferrite and Austenite Phases on the Overall Pitting Behavior of Lean Duplex Stainless Steels in Thiosulfate-Containing Environments. *J. Electrochem. Soc.* **2020**, *167*, 41502. [CrossRef]
10. Zhang, J.; Zhu, Y.; Xi, X.; Xiao, Z. Altered microstructure characteristics and enhanced corrosion resistance of UNS S32750 duplex stainless steel via ultrasonic surface rolling process. *J. Mater. Process. Technol.* **2022**, *309*, 117750. [CrossRef]
11. Zhang, Y.; Cheng, F.; Wu, S. Improvement of pitting corrosion resistance of wire arc additive manufactured duplex stainless steel through post-manufacturing heat-treatment. *Mater. Charact.* **2021**, *171*, 110743. [CrossRef]
12. Wang, T.; Phelan, D.; Wexler, D.; Qiu, Z.; Cui, S.; Franklin, M.; Guo, L.; Li, H. New insights of the nucleation and subsequent phase transformation in duplex stainless steel. *Mater. Charact.* **2023**, *203*, 113115. [CrossRef]
13. Zhang, Y.; Wang, C.; Reddy, K.M.; Li, W.; Wang, X. Study on the deformation mechanism of a high-nitrogen duplex stainless steel with excellent mechanical properties originated from bimodal grain design. *Acta Mater.* **2022**, *226*, 117670. [CrossRef]
14. Shang, F.; Chen, X.; Zhang, P.; Ji, Z.; Ming, F.; Ren, S.; Qu, X. Novel Ferritic Stainless Steel with Advanced Mechanical Properties and Significant Magnetic Responses Processed by Selective Laser Melting. *Mater. Trans.* **2019**, *60*, 1096–1102.
15. Roos, S.; Botero, C.; Rännar, L. Electron beam powder bed fusion processing of 2507 super duplex stainless steel. as-built phase composition and microstructural properties. *J. Mater. Res. Technol.* **2023**, *24*, 6473–6483. [CrossRef]
16. Koukolíková, M.; Podaný, P.; Rzepa, S.; Brázda, M.; Kocijan, A. Additive manufacturing multi-material components of SAF 2507 duplex steel and 15-5 PH martensitic stainless steel. *J. Manuf. Process.* **2023**, *102*, 330–339. [CrossRef]
17. Zhang, D.; Liu, A.; Yin, B.; Wen, P. Additive manufacturing of duplex stainless steels—A critical review. *J. Manuf. Process.* **2022**, *73*, 496–517. [CrossRef]
18. Jiang, D.; Gao, X.; Zhu, Y.; Hutchinson, C.; Huang, A. In-situ duplex structure formation and high tensile strength of super duplex stainless steel produced by directed laser deposition. *Mater. Sci. Eng. A* **2022**, *833*, 142557. [CrossRef]
19. Chen, L.; Liang, S.; Liu, Y.; Zhang, L. Additive manufacturing of metallic lattice structures: Unconstrained design, accurate fabrication, fascinated performances, and challenges. *Mater. Sci. Eng. R Rep.* **2021**, *146*, 100648. [CrossRef]
20. Haghdadi, N.; Laleh, M.; Moyle, M.; Primig, S. Additive manufacturing of steels: A review of achievements and challenges. *J. Mater. Sci.* **2021**, *56*, 64–107. [CrossRef]
21. Köhler, M.L.; Kunz, J.; Herzog, S.; Kaletsch, A.; Broeckmann, C. Microstructure analysis of novel LPBF-processed duplex stainless steels correlated to their mechanical and corrosion properties. *Mater. Sci. Eng. A Struct. Mater. Prop. Microstruct. Process.* **2021**, *801*, 140432. [CrossRef]
22. Liu, G.; Zhang, X.; Chen, X.; He, Y.; Cheng, L.; Huo, M.; Yin, J.; Hao, F.; Chen, S.; Wang, P.; et al. Additive manufacturing of structural materials. *Mater. Sci. Eng. R Rep.* **2021**, *145*, 100596. [CrossRef]
23. Gor, M.; Soni, H.; Wankhede, V.; Sahlot, P.; Grzelak, K.; Szachgluchowicz, I.; Kluczyński, J. A Critical Review on Effect of Process Parameters on Mechanical and Microstructural Properties of Powder-Bed Fusion Additive Manufacturing of SS316L. *Materials* **2021**, *14*, 6527. [CrossRef] [PubMed]
24. Astafurov, S.; Astafurova, E. Phase Composition of Austenitic Stainless Steels in Additive Manufacturing: A Review. *Metals* **2021**, *11*, 1052. [CrossRef]

25. Coldsnow, K.; Yan, D.; Paul, G.E.; Torbati-Sarraf, H.; Poorganji, B.; Ertorer, O.; Tan, K.; Pasebani, S.; Torbati-Sarraf, S.A.; Isgor, O.B. Electrochemical behavior of alloy 22 processed by laser powder bed fusion (L-PBF) in simulated seawater and acidic aqueous environments. *Electrochim. Acta* **2022**, *421*, 140519. [CrossRef]
26. Shang, F.; Chen, X.; Wang, Z.; Ji, Z.; Ming, F.; Ren, S.; Qu, X. The Microstructure, Mechanical Properties, and Corrosion Resistance of UNS S32707 Hyper-Duplex Stainless Steel Processed by Selective Laser Melting. *Metals* **2019**, *9*, 1012. [CrossRef]
27. Davidson, K.P.; Singamneni, S.B. Metallographic evaluation of duplex stainless steel powders processed by selective laser melting. *Rapid Prototyp. J.* **2017**, *23*, 1146–1163. [CrossRef]
28. Huang, C.; Shih, C. Effects of nitrogen and high temperature aging on σ phase precipitation of duplex stainless steel. *Mater. Sci. Eng. A* **2005**, *402*, 66–75. [CrossRef]
29. Salvetr, P.; Školáková, A.; Melzer, D.; Brázda, M.; Duchoň, J.; Drahokoupil, J.; Svora, P.; Msallamová, Š.; Novák, P. Characterization of super duplex stainless steel SAF2507 deposited by directed energy deposition. *Mater. Sci. Eng. A* **2022**, *857*, 144084. [CrossRef]
30. Nigon, G.N.; Burkan Isgor, O.; Pasebani, S. The effect of annealing on the selective laser melting of 2205 duplex stainless steel: Microstructure, grain orientation, and manufacturing challenges. *Opt. Laser Technol.* **2021**, *134*, 106643. [CrossRef]
31. Kunz, J.; Boontanom, A.; Herzog, S.; Suwanpinij, P.; Kaletsch, A.; Broeckmann, C. Influence of hot isostatic pressing post-treatment on the microstructure and mechanical behavior of standard and super duplex stainless steel produced by laser powder bed fusion. *Mater. Sci. Eng. A* **2020**, *794*, 139806. [CrossRef]
32. Papula, S.; Song, M.; Pateras, A.; Chen, X.B.; Brandt, M.; Easton, M.; Yagodzinskyy, Y.; Virkkunen, I.; Hanninen, H. Selective Laser Melting of Duplex Stainless Steel 2205: Effect of Post-Processing Heat Treatment on Microstructure, Mechanical Properties, and Corrosion Resistance. *Materials* **2019**, *12*, 2468. [CrossRef] [PubMed]
33. Zhang, J.; Dong, H.; Xi, X.; Tang, H.; Li, X.; Rao, J.H.; Xiao, Z. Enhanced mechanical performance of duplex stainless steels via dense core-shell nano-inclusions in-situ formed upon selective laser melting. *Scr. Mater.* **2023**, *237*, 115711. [CrossRef]
34. Zhang, B.; Li, H.; Zhang, S.; Jiang, Z.; Lin, Y.; Feng, H.; Zhu, H. Effect of nitrogen on precipitation behavior of hyper duplex stainless steel S32707. *Mater. Charact.* **2021**, *175*, 111096. [CrossRef]
35. Saeidi, K.; Alvi, S.; Lofaj, F.; Petkov, V.I.; Akhtar, F. Advanced Mechanical Strength in Post Heat Treated SLM 2507 at Room and High Temperature Promoted by Hard/Ductile Sigma Precipitates. *Metals* **2019**, *9*, 199. [CrossRef]
36. Saeidi, K.; Kevetkova, L.; Lofaj, F.; Shen, Z. Novel ferritic stainless steel formed by laser melting from duplex stainless steel powder with advanced mechanical properties and high ductility. *Mater. Sci. Eng. A* **2016**, *665*, 59–65. [CrossRef]
37. Hengsbach, F.; Koppa, P.; Duschik, K.; Holzweissig, M.J.; Burns, M.; Nellesen, J.; Tillmann, W.; Tröster, T.; Hoyer, K.; Schaper, M. Duplex stainless steel fabricated by selective laser melting—Microstructural and mechanical properties. *Mater. Des.* **2017**, *133*, 136–142. [CrossRef]
38. Haghdadi, N.; Cizek, P.; Hodgson, P.D.; He, Y.; Sun, B.; Jonas, J.J.; Rohrer, G.S.; Beladi, H. New insights into the interface characteristics of a duplex stainless steel subjected to accelerated ferrite-to-austenite transformation. *J. Mater. Sci.* **2020**, *55*, 5322–5339. [CrossRef]
39. Pan, C.; Hu, X.; Lin, P.; Chou, K. Effects of Ti and Al addition on the Formation and Evolution of Inclusions in Fe-17Cr-9Ni Austenite Stainless Steel. *Metall. Mater. Trans. B Process Metall. Mater. Process. Sci.* **2020**, *51*, 3039–3050. [CrossRef]

 *materials*

Article

Printing, Debinding and Sintering of 15-5PH Stainless Steel Components by Fused Deposition Modeling Additive Manufacturing

Gaoyuan Chang [1], Xiaoxun Zhang [1,*], Fang Ma [2], Cheng Zhang [1] and Luyang Xu [1]

[1] School of Materials Science and Engineering, Shanghai University of Engineering Science, Shanghai 201620, China
[2] School of Mechanical and Automotive Engineering, Shanghai University of Engineering Science, Shanghai 201620, China
* Correspondence: xxzhangsh@163.com

Abstract: Metal FDM technology overcomes the problems of high cost, high energy consumption and high material requirements of traditional metal additive manufacturing by combining FDM and powder metallurgy and realizes the low-cost manufacturing of complex metal parts. In this work, 15-5PH stainless steel granules with a powder content of 90% and suitable for metal FDM were developed. The flowability and formability of the feedstock were investigated and the parts were printed. A two-step (solvent and thermal) debinding process is used to remove the binder from the green part. After being kept at 75 °C in cyclohexane for 24 h, the solvent debinding rate reached 98.7%. Following thermal debinding, the material's weight decreased by slightly more than 10%. Sintering was conducted at 1300 °C, 1375 °C and 1390 °C in a hydrogen atmosphere. The results show that the shrinkage of the sintered components in the X-Y-Z direction remains quite consistent, with values ranging from 13.26% to 19.58% between 1300 °C and 1390 °C. After sintering at 1390 °C, the material exhibited a relative density of 95.83%, a hardness of 101.63 HRBW and a remarkable tensile strength of 770 MPa. This work realizes the production of metal parts using 15-5PH granules' extrusion additive manufacturing, providing a method for the low-cost preparation of metal parts. And it provides a useful reference for the debinding and sintering process settings of metal FDM. In addition, it also enriches the selection range of materials for metal FDM.

Keywords: metal fused deposition modeling; additive manufacturing; 15-5PH stainless steel; debinding; sintering

check for updates

Citation: Chang, G.; Zhang, X.; Ma, F.; Zhang, C.; Xu, L. Printing, Debinding and Sintering of 15-5PH Stainless Steel Components by Fused Deposition Modeling Additive Manufacturing. *Materials* **2023**, *16*, 6372. https://doi.org/10.3390/ma16196372

Academic Editor: Federico Mazzucato

Received: 22 August 2023
Revised: 16 September 2023
Accepted: 18 September 2023
Published: 23 September 2023

1. Introduction

Additive manufacturing (AM) stands as a groundbreaking technology that utilizes computer-aided design models to construct products by adding material layer by layer. Unlike traditional manufacturing methods, AM allows for the creation of products with diverse materials, complex shapes, structures and functions, without the need for subsequent post-processing such as cutting and machining [1]. The common materials used in AM include metals, ceramics and plastics, and the choice of AM method varies depending on the specific material being used [2]. At present, predominant metal AM techniques encompass selective laser melting (SLM) [3–5], selective laser sintering (SLS) [6,7], electron beam melting (EBM) [8,9], direct laser metal sintering (DLMF) [10], binder injection (BJ) [11], wire arc additive manufacturing (WAAM) [12] and laser bed powder fusion (LBPF) [13]. Most of these techniques use lasers to subject the material to a process of high temperature and rapid cooling. However, the high temperature gradients and rapid cooling under these processes cause material anisotropy and generate high residual stresses that affect the mechanical properties of the material [14–17]. In addition, it is too expensive to purchase this equipment, with high maintenance and repair costs, which require a great quantity of upfront investment. For example, SLM and EBM printers require laser or electron beams and inert or vacuum chambers, and WAAM technology preparation requires high-precision equipment, which is complex and expensive. Therefore, developing a low-cost metal AM method has great practical significance. Fused deposition molding (FDM) emerges as a promising low-cost AM technology that operates at room temperature. The material at the nozzle is heated to the melting temperature (typically between 200 °C and 300 °C), without complex energy sources. This makes FDM equipment much cheaper than equipment such as SLM. The process involves heating and melting a hot molten material, which is then extruded through a printing nozzle. The nozzle moves along a specific trajectory under computer control, depositing the material in a semi-fluid state onto a printing platform. The material solidifies and forms the final solid product by stacking layers. Another technique, known as metal injection molding (MIM), involves injecting a mixture of metal powder and polymer into a molding chamber to achieve the desired shape [18,19]. The polymer is then removed through a debinding process, followed by sintering the metal powder below its melting point to obtain densely consolidated metal parts. By incorporating the pellets and post-processing techniques used in MIM into FDM, it is possible to combine the advantages of both technologies. This allows for the production of complex structural parts using FDM, a low-cost technology, while achieving dense metal parts through the debinding and sintering processes. In this process, the printed part is often referred to as the "green part", the solvent debinded part is called the "brown part", and the sintered part is called the "FDMS part". The feedstock used in metal FDM technology is a mixture of metal powder and a specified proportion of polymer material, usually in the form of 1.75 mm diameter wire or approximately 3 mm diameter pellets. The fluidity of the material is provided by the polymer in the feedstock, which only needs to be heated at the nozzle for smooth extrusion. Metal powders and polymers are readily available, and the pellet feedstock for this process can be prepared at a cost of as little as 20 USD/kg. The printing, debinding and sintering equipment used in the metal FDM is cheaper and less expensive to maintain than those requiring electronics and lasers. These features of metal FDM significantly reduce the cost of producing metal parts by AM, making it suitable for large-scale use in laboratories and small businesses.

Many scholars have shown great interest in metal FDM in recent years. Yvonne Thompson et al. [20] prepared printable FFF filaments from grafted polyolefin, thermoplastic elastomer and 55 vol.% 316L powder and optimized printing parameters, particularly studying the debinding and sintering process, resulting in 316L specimens with a shrinkage of about 20% and a relative density of more than 95%. M. Sadaf [21] developed filaments containing 65 vol.% of 316L steel powder using a one-component binder, which avoided the solvent debinding process. This was followed by sintering the green parts at 1380 °C under hydrogen to obtain metal specimens with a tensile strength of 520 MPa and a hardness of 285 HV. Liu Bin et al. [22] prepared 316L/POM filaments and used catalytic debinding to remove the binder, studied the microstructural characteristics of the filaments, green parts and sintered parts and tested the relative density, tensile properties, hardness and shrinkage of the sintered parts. Gurminder Singh et al. [23] investigated commercial MIM Cu raw materials using screw-type printers. They delved into the impact of layer thickness, nozzle temperature, extrusion multiplicity and printing speed on the density and surface roughness of green parts. Utilizing a central composite design approach, they assessed these factors and employed micro-tomography to examine sample porosity under various process parameters. The optimised green parts were also sintered to obtain high-density sintered copper parts with a low surface roughness. A high metal powder content results in a low shrinkage of the sintered parts. The researchers therefore expect to be able to increase the metal powder content as much as possible while maintaining normal printing conditions. Materials with a relatively high metal powder content have a poor flowability, and preparing filaments that can pass through the printer rollers without breaking is a difficult task. Furthermore, the preparation of filaments is already a lengthy process compared to granules, prompting researchers to explore screw printers as an alternative solution.

Similarly, Gurminder Singh et al. [24] used the MIM17-4PH feedstock for printing under optimum parameters to investigate the density of the sintered parts at different temperatures. Qualitative and quantitative analysis of the pores in the green and sintered parts using micro-tomography confirmed that the optimized printing parameters were beneficial for the final microstructure. In addition, several researchers have also investigated the FDM process of other metallic or ceramic materials; for example, 316L [25–27], 17-4PH [28,29], copper [30,31], hardmetal [32], Ti6Al4V [33,34], tungsten heavy alloy [35], H13 [36], Al [37], 1.2083 steel [38], zirconia [39,40], etc. However, no studies on the FDM of 15-5PH stainless steel (15-5PH SS) materials were found in the available literature.

The 15-5PH stainless steel, derived from 17-4PH steel through a reduction in Cr content and increase in Ni content, represents a martensitic precipitation-hardening stainless steel renowned for its exceptional mechanical properties and corrosion resistance [41]. Therefore, it finds widespread use as an engineering material in aerospace, medical, chemical and other fields [42–44]. At the same time, 3D printing is gradually maturing, and the application scenarios are constantly expanding. In this work, we developed a 15-5PH granular feedstock with a metal powder content of up to 90 wt.% and successfully constructed parts using a screw-type printer and produced metal parts after debinding and sintering. Firstly, this research delved into the impact of flowability and the printing process parameters of 15-5PH granular material on the forming state of the parts. Furthermore, we explored both solvent and thermal debinding processes. Subsequently, this work investigated the influence of sintering temperature on the densities, shrinkage, mechanical properties and micromorphology of the sintered parts. Aiming at the current problems of metal AM with expensive equipment, complicated operation, and mainly oriented to high-precision enterprises, this study aims to explore a low-cost metal additive manufacturing technology that is simple to operate and easy to popularize. This work not only achieves low-cost metal AM but also expands the range of materials used for metal FDM.

2. Experimental Details

The process flow diagram of this work is depicted in Figure 1. Detailed elaboration of the specific steps will be provided in the subsequent sections.

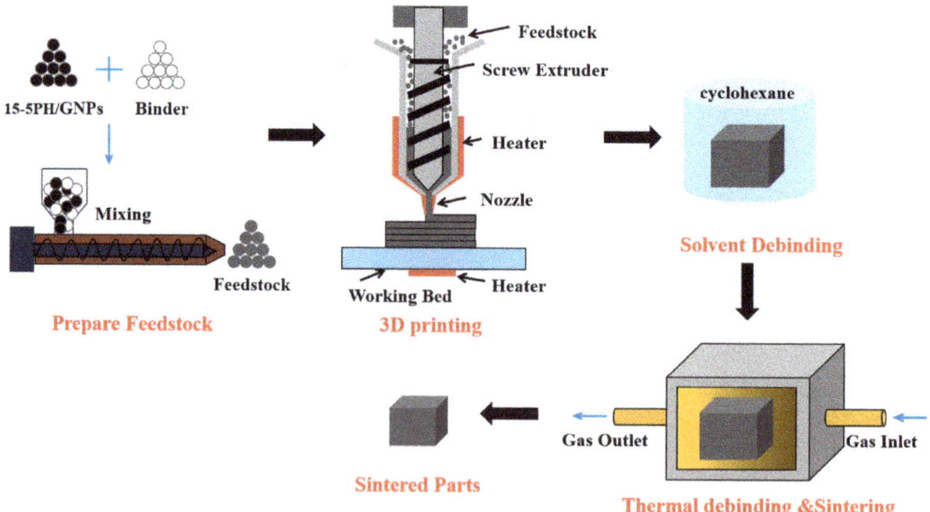

Figure 1. Process flow diagram of 15-5PH stainless steel by metal FDM.

2.1. Materials and Preparation of Feedstock

The metal powder employed in this investigation is a gas-atomized 15-5PH SS powder with a sphericity of 96%, supplied by PMG 3D Technologies (Shanghai) Co., Ltd. (Shanghai, China). The powder particle size is within the range of 15–45 µm. All illustration of the powder's morphology is presented in Figure 2. The chemical composition of 15-5PH SS is provided by the distributor as shown in Table 1. Thermoplastic elastomer (TPE) was selected as the soluble binder and maleic anhydride grafted polypropylene (MAH-g-PP) as the insoluble backbone. Both binders were purchased from Dongguan Qihong Plastic Co., Ltd. (Dongguan, China). The MAH-g-PP was produced by introducing a strong polar maleic anhydride side branch into the main chain of the non-polar molecules of polypropylene, which can enhance the compatibility of polar and non-polar materials and facilitate the dispersion of metal powders. None of the above raw materials were further treated before the experiment. The proportions of metal powders and binder compositions used in this work are given in Table 2.

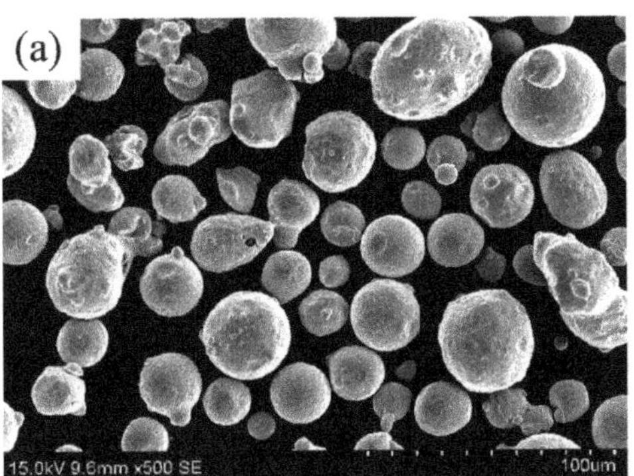

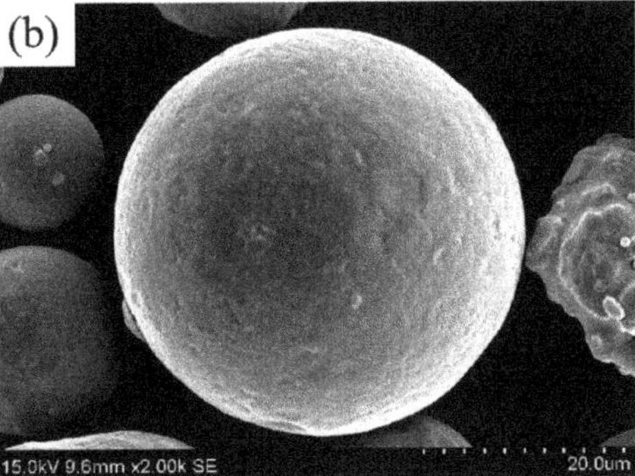

Figure 2. SEM image of 15-5PH stainless steel powder: (**a**) with magnifying powder of 500×, (**b**) with magnifying powder of 2000×.

Table 1. Chemical compositions of 15-5PH SS powders versus national standards (wt%).

Elements	C	Si	Mn	S	P	Cr	Ni	Cu	Nb	Fe
ASTM	≤0.07	≤1.00	≤1.00	≤0.015	≤0.03	14.0–15.5	3.5–5.5	2.5–4.5	0.15–0.45	Bal
15-5PH	0.015	0.59	0.52	0.004	0.03	15.22	3.98	3.85	0.35	Bal

Table 2. Composition and proportion of 15-5PH granular feedstock.

Feedstock	15-5PH Powder	TPE	MAH-g-PP
Content (wt.%)	90	7	3

The 15-5PH granules that can be used for 3D printing were prepared as follows: 15-5PH stainless steel powder and the binder were put into a high-speed mixer for mixing according to the ratio in the Table 2. The mixed material was fed by the conical twin-screw extruder for compound extrusion. The twin-screw extruder was maintained at temperatures of 190 °C, 195 °C, 195 °C, and 200 °C in each respective zone, while the extrusion speed was set as 25 rpm. The extruded material was subsequently granulated using a granulator. To ensure the even distribution of the metal powder in the binder, the granules were again extruded in the extruder. The reextruded particles are placed in a vacuum drying oven for 3 h and stored in a vacuum environment for later printing.

2.2. 3D Printing

The part was initially designed in Solidworks (version number: 2021) software and then converted to the STL format. Subsequently, the model underwent slicing using Simplify3D software (version number: 4.0.1) and was imported into a G5 Pro printer from Shenzhen Piocreat 3D Technology Co., Ltd. (Shenzhen, China). The printer is a screw-type printer, which can print granular materials directly, avoiding the process of filament preparation. Its morphology and structure are shown in Figure 3a,b. The printer transports particles through a screw movement to a heating unit, where the material is melted and extruded through nozzles to be deposited on the platform. The printer platform is a metal substrate with a good thermal conductivity that heats up quickly. The machine has a print size of 500 × 500 × 500 mm and a print speed range of 0–100 mm/s. The slice thickness is 0.1–1 mm and the positioning accuracy of the X-Y axis is ±0.1 mm, allowing the production of complex geometries. In this study, dog bone tensile specimens (length 66 mm, thickness 3 mm, maximum width 12 mm, minimum width 4 mm) and small blocks (15 mm × 15 mm × 3 mm) were printed. Figure 3c presents the morphology of the printed, solvent debinded and sintered part. To match the material's flow characteristics, a nozzle diameter of 0.8 mm was selected. The printing parameters were set as follows: fill rate 100%, flow rate multiplier 180%, layer thickness 0.2 mm, printing speed 40 mm/s, nozzle temperature 285 °C, platform temperature 90 °C and fill angle ±45°. The nozzle fan was switched off during the printing process.

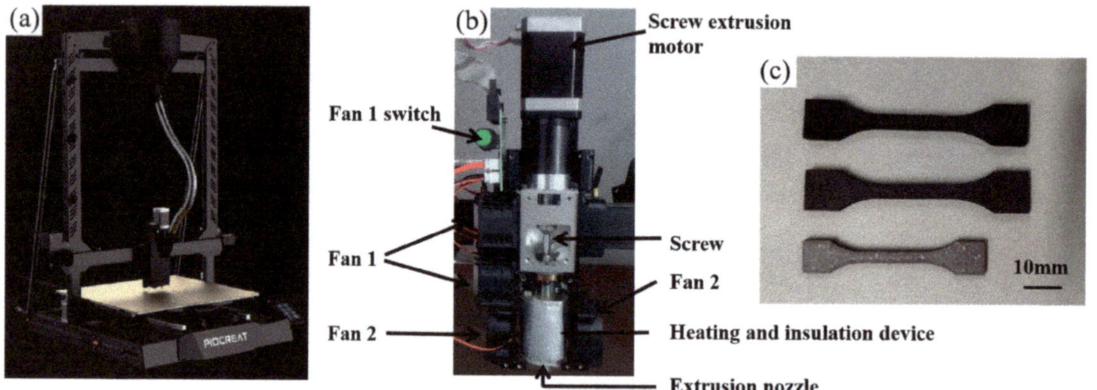

Figure 3. (a) G5Pro printer, (b) the structure of G5Pro printer and (c) the parts after being printed (up), debinded (middle) and sintered (down).

2.3. Debinding and Sintering

The debinding process consists of solvent debinding and thermal debinding. Solvent debinding removes the soluble binder TPE. The parts were immersed in cyclohexane at 25 °C, 45 °C, 65 °C and 75 °C for 24 h while being palced in a constant temperature magnetic stirrer. The parts were removed every two hours, then dried at 80 °C for 3 h within a vacuum oven before weighed. The rate of binder removal during solvent debinding was evaluated by calculating the ratio of binder loss after solvent debinding to the total binder. The ratio of weight loss in each step to the total weight was used to evaluate the binder removal rate throughout the debinding process.

Thermal debinding and sintering were conducted in a tube furnace (Shanghai Shiheng Instrument Equipment Co., Ltd. (Shanghai, China)), where the samples were heated to 1300 °C, 1375 °C and 1390 °C, respectively, and held for 3 h. Hydrogen gas was introduced at a flow rate of 250 mL/min. Thermal debinding removes the backbone binder from the sample and ends the debinding process. To establish the thermal debinding curve, we performed thermogravimetric analysis (TGA) on the granules. The TGA results, depicted in Figure 4a, were obtained in a nitrogen atmosphere with a heating rate of 10 °C/min, ranging from 30 °C to 700 °C. The results showed that the material started to decompose at 365 °C and finished at 471 °C. The green parts are heated to 600 °C and held at 350 °C and 600 °C to ensure complete binder removal during the thermal debinding. The heating rates applied for both thermal debinding and sintering were 2.0 °C/min and 2.5 °C/min, respectively. After insulation at the sintering temperature, the samples were cooled within the furnace. The corresponding thermal debinding and sintering curves are shown in Figure 4b.

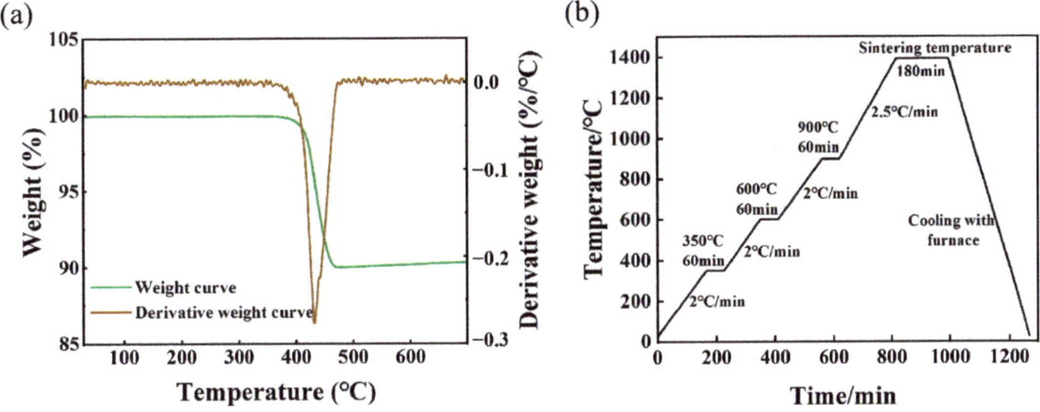

Figure 4. (**a**) TGA of 15-5PH granular feedstock; (**b**) thermal debinding and sintering curves.

2.4. Characterization and Test

In this study, an S-3400N scanning electron microscope (SEM) was used to observe the morphology of the granules, green parts, brown parts and sintered parts. The flowability of 15-5PH granules was tested at 265 °C, 275 °C, 285 °C and 295 °C using a capillary rheometer with a load of 21.6 kg. For the determination of sintered part density, we utilized Archimedes' principle, repeating each measurement three times to ensure experimental precision. The sintered parts were corroded with aqua regia to observe the grain morphology and porosity. To calculate the shrinkage of the parts, the dimensions of the parts before and after sintering were measured with vernier calipers. Furthermore, the mechanical properties were evaluated through hardness and tensile tests on the sintered components, aiming to investigate the influence of sintering temperature. Sintered parts were subjected to hardness assessment using a Rockwell hardness tester, with ten measurements taken per sample, and the average value determined as the final hardness value. Tensile tests were conducted at room temperature, employing an AG-25TA tensile tester operating at a speed of 1 mm/min. Three tensile specimens were prepared under each sintering temperature to reduce the experimental error, and the results were averaged.

3. Results and Discussion

3.1. Feedstock Characterization and 3D Printing

The microscopic morphology of the 15-5PH granular feedstock is depicted in Figure 5. Figure 5a,b illustrate that the metal powders are uniformly distributed between the binders without agglomeration. The homogeneous mixing of the metal powder and binder can prevent visible defects in the sintered parts and avoid warping and cracking [45]. The printing quality is closely related to the flowability of the feedstock [20,23]. Melt mass flow rate (MFR) measurements were performed on the granules to determine the nozzle temperature, and the results are presented in Figure 5c. The result indicates that the MFR of the feedstock exhibits sensitivity to temperature variations. As the temperature rises, the MFR tends to increase and then decrease, reaching the optimum value at 285 °C. Practice has also shown that the material extrusion process is extremely smooth when the nozzle temperature is 285 °C.

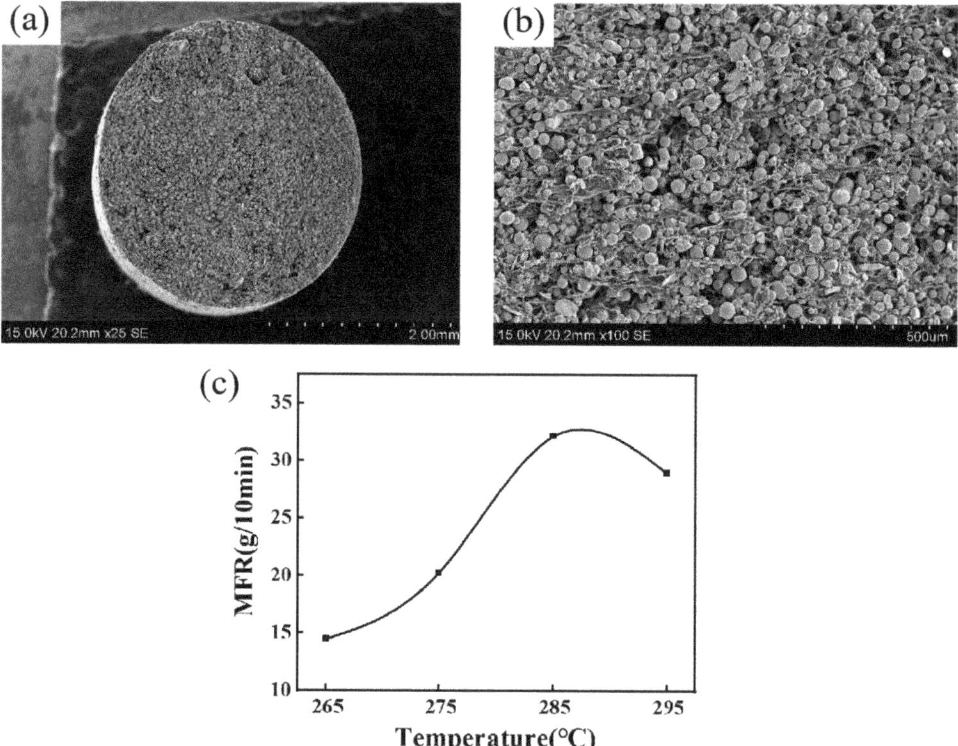

Figure 5. SEM image of 15-5PH granular feedstock: (**a**) 25×, (**b**) 100×; (**c**) melt mass flow rate of 15-5PH granular feedstock.

The 0.6 mm diameter nozzle often clogged and could not print normally. After changing the nozzle diameter to 0.8 mm, the printer extruded smoothly. The results of matching the flowability of the feedstock to the printer showed that when the extrusion multiplier was 100%, the material was under-extruded, resulting in parts that could not be moulded. Adjusting the extrusion multiplier to 180% resulted in a good part-molding quality. It was found that high-surface-quality parts could be obtained by switching off the fan at the nozzle during the printing process.

Figure 6 presents the microscopic topography of the green part. Figure 6a,b show that the metal powder is still uniformly distributed in the binder after printing with the screw-type printer. Figure 6c,d show the layer thickness of the green part and the interlayer line during interlayer bonding, which is caused by the characteristics of FDM. The formation of interlayer lines introduces pores into the parts, which can impact the densities and mechanical properties of the sintered parts. The interlayer bonding effect can be modified by adjusting the printing parameters, thereby changing the mechanical properties of the parts [23,29,46].

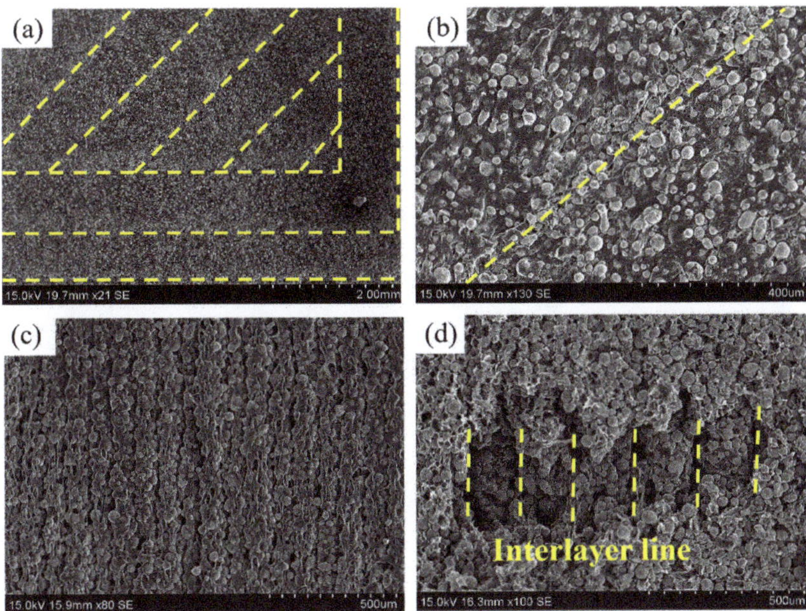

Figure 6. (**a,b**) Microscopic morphology of the green part, (**c,d**) side surface of the green part with magnifications of 80× and 100×.

3.2. Solvent and Thermal Debinding

TPE is soluble in cyclohexane, and solvent debinding is used to remove soluble binders from the print [26]. The green parts were immersed in cyclohexane at different temperatures and kept for different times to investigate the effect of the debinding temperature and time on the solvent debinding rate, and the results are shown in Figure 7. The debinding rate increment of the parts decreased continuously over time at four temperatures. On the contrary, the debinding rate increased continuously over time, and the change in debinding rate was not significant after 14 h. The solvent debinding process mainly involves dissolution and diffusion, and the concentration of the solution and the distance of the solution diffusion path will affect the debinding rate increment. Figure 8 is a schematic diagram of the solvent debinding process. At the initial stage of solvent debinding (Figure 8a), the solution is in contact with the surface of the parts, the diffusion path is the shortest, and the solution concentration is the highest. As shown in Figure 7, the debinding rate increases from 0 to 57.83% in 2 h at a temperature of 75 °C. In the middle stage of solvent debinding, both the solution concentration and the diffusion path increase with time, as shown in Figure 8b, leading to a reduction in the debinding rate increment, although the debinding rate continues to rise. By the final stage of solvent debinding, the soluble binder is almost completely removed. Only PP remains in the printed parts, maintaining their shape, as shown in Figure 8c. The debinding rate increases with the increasing temperature for the same time, because the dissolution and diffusion velocity increase with increasing temperature. In addition, the solubility of cyclohexane increases at higher temperatures, which can dissolve more soluble components. The samples after solvent debinding (Figure 9) showed that pores had formed inside the parts and connecting holes had formed in some areas. These pores will provide channels for gas to escape during the thermal debinding process, preventing the bulging and cracking of the part [20].

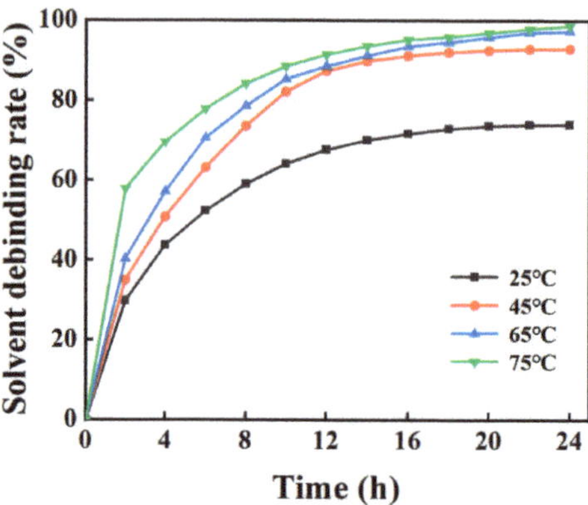

Figure 7. Solvent debinding rate for the green part at different temperatures and times.

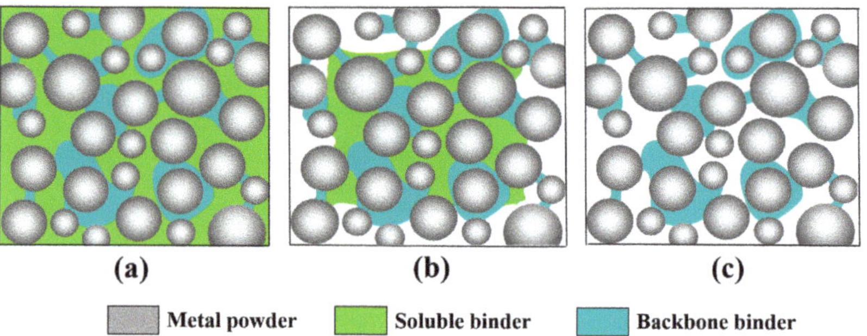

Figure 8. Schematic diagram of solvent debinding process. (**a**) the initial stage of solvent debinding, (**b**) the middle stage of solvent debinding; (**c**) the final stage of solvent debinding.

Figure 9. Microstructure of green part: (**a**,**b**) before solvent debinding, (**c**,**d**) after solvent debinding.

The strength of the samples after thermal debinding proved to be insufficient for testing purposes. Therefore, the brown part was heated to 900 °C for pre-sintering to investigate the thermal debinding process. Figure 10 presents the morphology of the part after this process. After thermal debinding and pre-sintering, the binder has completely disappeared from the brown part and a sintering neck is formed between most of the metal particles. Table 3 provides the results of the total weight loss observed in the green parts after debinding. The binder mass in the granules is 10% of the total mass of the feedstock. The soluble components account for 7% and the backbone components account for 3% of the binder. Table 3 reveals that solvent debinding successfully removed the majority of the soluble binder from the green parts. Furthermore, after thermal debinding, both the backbone and any residual soluble binder were entirely eliminated. After debinding, the metal particles on the surface of the sample were lost during transport, resulting in a total weight loss of just over 10% of the feedstock.

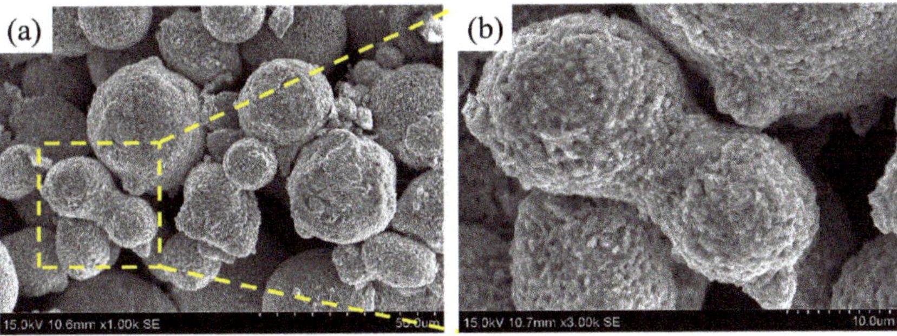

Figure 10. SEM morphology of the parts after thermal debinding and pre-sintering. (**a**) the sintering neck between the metal powder, (**b**) left image magnified 3000×.

Table 3. Mass change after different debinding steps relative to the mass of green part.

Green Part mass/g	Solvent Debinding/(Δm,%)	Thermal Debinding/(Δm,%)	Total/(Δm,%)
10.7822	6.91%	3.16%	10.07%
10.7069	6.91%	3.19%	10.10%
10.6765	6.86%	3.17%	10.03%

3.3. Sintered Parts: Shrinkage, Relative Density, Microstructure and Mechanical Properties

The brown part undergoes high-temperature sintering to yield a densely consolidated metal component. Notably, brown parts sintered at different temperatures exhibit an excellent appearance, devoid of defects such as warping, collapse, bulging and cracking. Figure 11a illustrates the shrinkage observed in brown parts after sintering at 1300 °C, 1375 °C and 1390 °C for 3 h. The shrinkage of the sintered parts in the X-Y-Z direction remains relatively consistent at the same sintering temperature. The model can be scaled up accordingly during slicing to compensate for the dimensional shrinkage that occurs after sintering. As the temperature increases, the atomic migration velocity and the growth velocity of sintering necks increase, and the grain boundary growth velocity across the pores increases. Consequently, the internal pore volume within the part decreases compared to low-temperature sintering, ultimately resulting in an increase in dimensional shrinkage for the sintered part as the temperature rises.

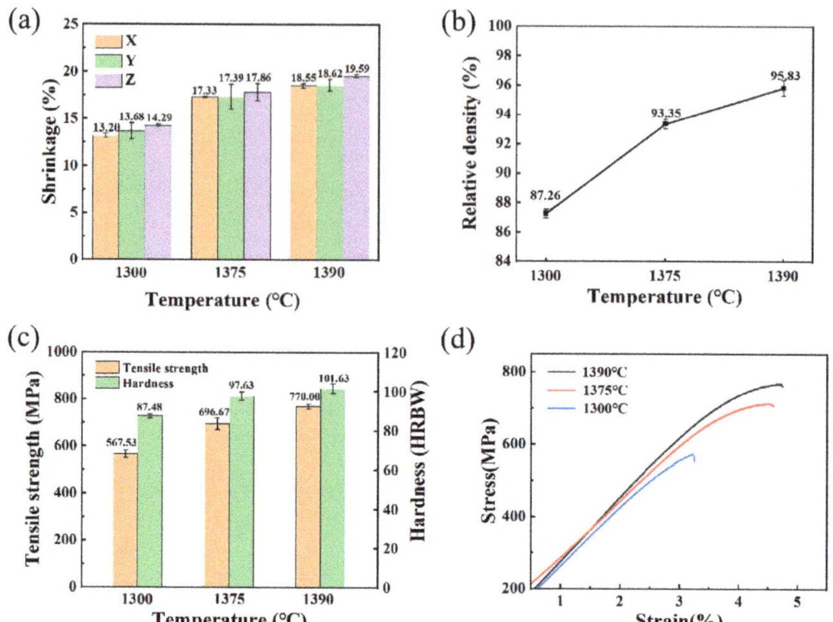

Figure 11. (**a**) Effect of sintering temperature on dimensional shrinkage, (**b**) effect of sintering temperature on relative density, (**c**) effect of sintering temperature on tensile strength and hardness, (**d**) stress–strain curve of the sintered parts.

The relationship between relative density and sintering temperature is illustrated in Figure 11b. As the sintering temperature rises, there is a corresponding increase in the relative density of the sintered part, mirroring the trend observed in dimensional shrinkage with temperature. The relative density of the sintered part reached 95.83% at 1390 °C, a figure comparable to that of sintered parts produced via MIM [47].

Figure 12 provides the microstructure of sintered parts at various temperature levels. When sintered at 1300 °C (Figure 12a), the number of pores in sintered parts was high and the size of the pores was large. Although the sintering neck between the powders grew with increasing temperatures between 900 °C and 1300 °C, the density was still low. There was no obvious grain morphology inside the parts after aqua regia corrosion. Following sintering at 1375 °C, there was a marked reduction in the number of pores within the samples, and the remaining pores tended to be mostly spherical and smaller in size. It can be seen that the structure is martensite after corrosion. With a further increase in temperature from 1375 °C to 1390 °C, both the number and size of pores continued to decrease. However, Figure 12c shows a increase in grain size.

The trend of the effect of sintering temperature on hardness and tensile strength is shown in Figure 11c. Both hardness and tensile strength increase with increasing sintering temperature, which is consistent with the trend of density with sintering temperature. Rockwell hardness is measured by pressing an indenter into the surface of an object with a certain force and determining the hardness from the depth of the indentation residue. The part sintered at 1300 °C contained a 12.74% porosity, which was easier for the indenter to press into during the hardness test, leaving deeper craters in the samples, resulting in a relatively low hardness of 87.48 ± 1.03 HRBW. After sintering at 1390 °C, the parts had a porosity of 4.17% and a high density. The dense structure impedes the indenter during the testing process, resulting in a shallower indentation depth remaining on the surface and a corresponding increase in hardness.

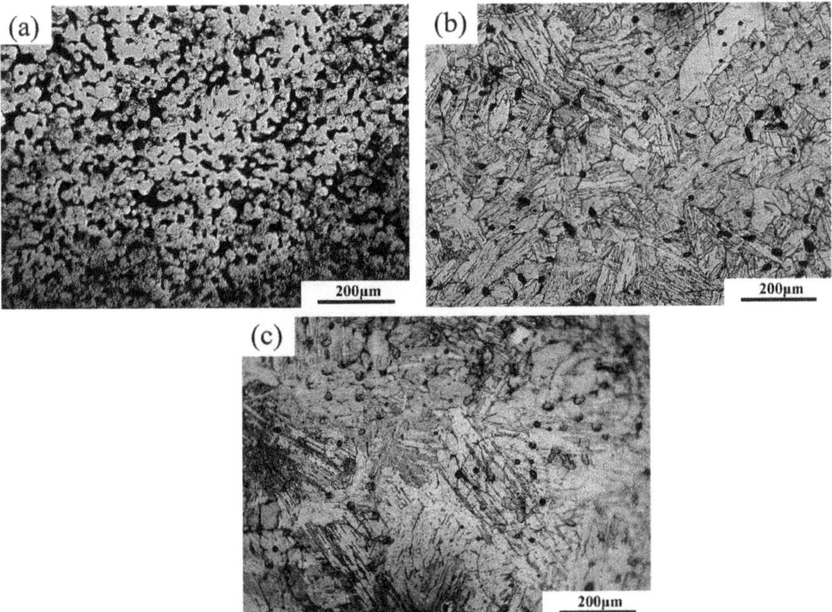

Figure 12. Microstructure of the parts sintered at (**a**) 1300 °C, (**b**) 1375 °C, (**c**) 1390 °C.

The tensile strength results presented in Figure 11c reveal a noteworthy increase in tensile strength, rising from 567.53 MPa to 770 MPa, representing a 35.68% increase as temperature escalates from 1300 °C to 1390 °C. Figure 11d shows the stress–strain curves of the sintered parts at different temperatures. These curves indicate that the 15-5PH steel parts experienced fracturing with a relatively small plastic deformation. Tensile strength and relative density are closely related. The part sintered at 1300 °C contains more pores, which act as a source of cracks during the tensile process, making the material unable to withstand too high of a load before fracture. As the temperature rises, the internal pores of the part gradually shrink or even close and the porosity decreases. The degree of pore spheroidisation is high and mostly small, so these pores require higher stress conditions to become crack sources, leading to the specimen being able to withstand higher loads. The fracture morphology of sintered parts at different temperatures is shown in Figure 13. Figure 13a shows a sample sintered at 1300 °C with many river-like patterns on the fracture surface, which is a typical appearance of a brittle fracture. The fracture morphology of the sintered part at 1390 °C (Figure 13b) shows that it still contains internal pores. It is noteworthy that in addition to the brittle fracture characteristics, there are also a large number of dimple-like micropores on the fracture surface at 1390 °C. Local magnification of the sintered parts at 1390 °C shows the presence of unmelted powders in the dimple-like micropores. These powders detach from the matrix under external forces to form micropores and gradually grow to form dimples.

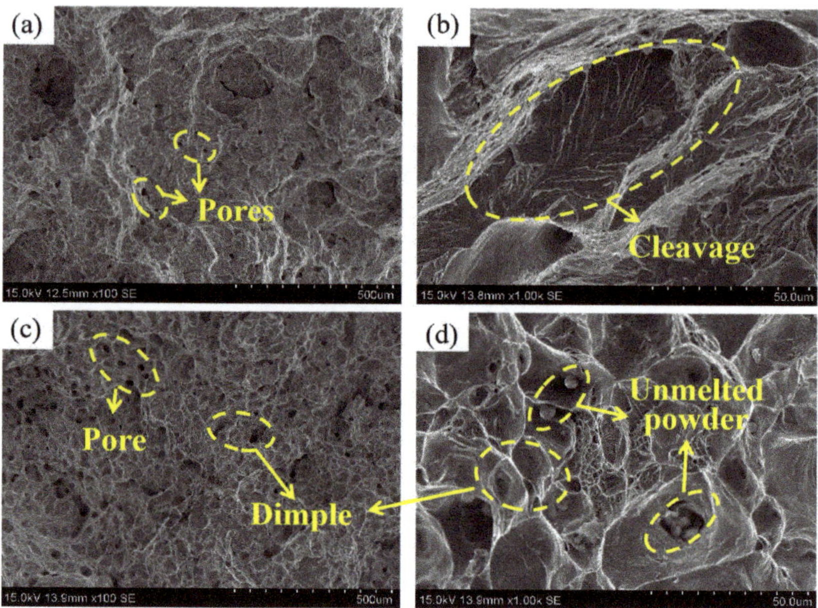

Figure 13. SEM images of the tensile fracture of sintered parts at different temperatures: (**a**,**b**) 1300 °C, (**c**,**d**) 1390 °C.

4. Conclusions

This work has successfully developed 15-5PH stainless steel granules that can be used for metal FDM. Printing, debinding and sintering processes of metal FDM were investigated using these granules as a raw material. Metal parts were fabricated using low-cost 3D printing technology. The main conclusions of the research are as follows:

1. The 15-5PH stainless steel powder is evenly distributed in the granules without agglomeration. The MFR of the granule materials is sensitive to temperature changes, and the fluidity of the granules is the best at 285 °C. The selection of the nozzle diameter and the adaptability of the printer to the viscosity of granules are key to successful printing. The condition of the fan at the nozzle determines the surface quality of the part. The optimum printing parameters are a nozzle diameter 0.8 mm, an extrusion multiplier 180% and the fan shut off at the nozzle.

2. Solvent debinding removes soluble components from green parts and provides a pathway for gas diffusion during the thermal debinding process. The solvent debinding rate increases continuously with increasing debinding temperature and time. The debinding rate reaches its maximum at a temperature of 75 °C for 24 h, which is 98.7%. At the same temperature, the debinding rate increment gradually decreases and eventually stabilizes over time. During the thermal debinding process, sintering necks form between the metal powders, preserving the part's structural integrity. All of the binder was removed and the weight loss was about 10% after debinding.

3. The relative density of sintered parts experiences a steady rise with increasing sintering temperature, progressing from 87.26% at 1300 °C to 95.83% at 1390 °C. The microstructure indicates that the parts sintered at 1300 °C contain many pores with large sizes. And the number and size of pores decrease significantly at 1390 °C. The dimensional shrinkage of the sintered parts remains uniform in the X-Y-Z directions. The shrinkage amplifies as the sintering temperature rises, with the range of shrinkage varying from 13.26% to 19.58% within the temperature range of 1300 °C to 1390 °C.

4. The hardness and tensile strength of sintered parts increase with increasing temperature, which is mainly related to the density of the part. The hardness of the sintered parts is 87.48 ± 1.03 HRBW at 1300 °C, and it does not change significantly between 1375 °C and 1390 °C. The tensile strength of the sintered parts increases from 567.53 MPa at 1300 °C to 770 MPa at 1390 °C, an increase of 35.68%. The fracture surface of the sintered parts at 1300 °C shows many dissociated sections, while the fracture surface at 1390 °C shows many dimples. The 15-5PH steel parts show brittle fracture with almost no plastic deformation.

This work achieves the low-cost additive manufacturing of metal parts, which has great advantages in the production of small batches and personalised parts. However, the mechanical properties of parts produced by metal FDM need to be improved, and better mechanical properties are expected to be achieved by adjusting the FDM printing parameters and optimising the debinding and sintering processes. Future research directions could focus on investigating printing parameters and changing sintering methods, such as using vacuum hot pressing sintering, microwave sintering and other sintering methods to improve the mechanical properties of materials. The sintered part can also be subjected to heat treatments such as solution and aging to improve the mechanical properties and expand its application scenarios.

Author Contributions: Conceptualization, G.C., C.Z. and L.X.; formal analysis, G.C.; funding acquisition, X.Z. and F.M.; investigation, G.C.; methodology, G.C., X.Z., C.Z. and L.X.; project administration, X.Z.; resources, X.Z.; supervision, X.Z., F.M., C.Z. and L.X.; writing—original draft, G.C.; writing—review and editing, G.C. All authors have read and agreed to the published version of the manuscript.

Funding: This research was funded by the Class III Peak Discipline of Shanghai-Materials Science and Engineering (High-Energy Beam Intelligent Processing and Green Manufacturing).

Informed Consent Statement: The authors declare that they have no known competing financialinterests or personal relationships that could have appeared to influence the work reported in thispaper.

Data Availability Statement: The data in this work are available from the corresponding authors on resonable request.

Acknowledgments: The authors thank the Shenzhen Piocreat 3D Technology Co., Ltd. (Shenzhen, China) for providing the printing equipment. The authors would like to thank the Shanghai Shiheng Instrument Equipment Co., Ltd. (Shanghai, China), for providing the sintering equipment.

Conflicts of Interest: The authors declare no conflict of interest.

References

1. Nurhudan, A.I.; Supriadi, S.; Whulanza, Y.; Saragih, A.S. Additive manufacturing of metallic based on extrusion process: A review. *J. Manuf. Process.* **2021**, *66*, 228–237. [CrossRef]
2. Gonzalez-Gutierrez, J.; Cano, S.; Schuschnigg, S.; Kukla, C.; Sapkota, J.; Holzer, C. Additive Manufacturing of Metallic and Ceramic Components by the Material Extrusion of Highly-Filled Polymers: A Review and Future Perspectives. *Materials* **2018**, *11*, 840. [CrossRef]
3. Han, Y.; Zhang, Y.; Jing, H.; Lin, D.; Zhao, L.; Xu, L.; Xin, P. Selective laser melting of low-content graphene nanoplatelets reinforced 316L austenitic stainless steel matrix: Strength enhancement without affecting ductility. *Addit. Manuf.* **2020**, *34*, 101381. [CrossRef]
4. Dong, M.; Zhou, W.; Kamata, K.; Nomura, N. Microstructure and mechanical property of graphene oxide/AlSi10Mg composites fabricated by laser additive manufacturing. *Mater. Charact.* **2020**, *170*, 110678. [CrossRef]
5. Ahmadi, M.; Tabary, S.B.; Rahmatabadi, D.; Ebrahimi, M.; Abrinia, K.; Hashemi, R. Review of selective laser melting of magnesium alloys: Advantages, microstructure and mechanical characterizations, defects, challenges, and applications. *J. Mater. Res. Technol.* **2022**, *19*, 1537–1562. [CrossRef]
6. Tong, Q.; Xue, K.; Wang, T.; Yao, S. Laser sintering and invalidating composite scan for improving tensile strength and accuracy of SLS parts. *J. Manuf. Process.* **2020**, *56*, 1–11. [CrossRef]
7. Olakanmi, E.O.T.; Cochrane, R.F.; Dalgarno, K.W. A review on selective laser sintering/melting (SLS/SLM) of aluminium alloy powders: Processing, microstructure, and properties. *Prog. Mater. Sci.* **2015**, *74*, 401–477. [CrossRef]

8. Murr, L.E.; Gaytan, S.M.; Ramirez, D.A.; Martinez, E.; Hernandez, J.; Amato, K.N.; Shindo, P.W.; Medina, F.R.; Wicker, R.B. Metal Fabrication by Additive Manufacturing Using Laser and Electron Beam Melting Technologies. *J. Mater. Sci. Technol.* **2012**, *28*, 1–14. [CrossRef]

9. Gong, H.; Snelling, D.; Kardel, K.; Carrano, A. Comparison of Stainless Steel 316L Parts Made by FDM- and SLM-Based Additive Manufacturing Processes. *JOM* **2018**, *71*, 880–885. [CrossRef]

10. Santos, E.C.; Shiomi, M.; Osakada, K.; Laoui, T. Rapid manufacturing of metal components by laser forming. *Int. J. Mach. Tools Manuf.* **2006**, *46*, 1459–1468. [CrossRef]

11. Huber, D.; Vogel, L.; Fischer, A. The effects of sintering temperature and hold time on densification, mechanical properties and microstructural characteristics of binder jet 3D printed 17-4 PH stainless steel. *Addit. Manuf.* **2021**, *46*, 14. [CrossRef]

12. Wang, J.; Pan, Z.; Wang, Y.; Wang, L.; Su, L.; Cuiuri, D.; Zhao, Y.; Li, H. Evolution of crystallographic orientation, precipitation, phase transformation and mechanical properties realized by enhancing deposition current for dual-wire arc additive manufactured Ni-rich NiTi alloy. *Addit. Manuf.* **2020**, *34*, 101240. [CrossRef]

13. Azami, M.; Siahsarani, A.; Hadian, A.; Kazemi, Z.; Rahmatabadi, D.; Kashani-Bozorg, S.F.; Abrinia, K. Laser powder bed fusion of Alumina/Fe–Ni ceramic matrix particulate composites impregnated with a polymeric resin. *J. Mater. Res. Technol.* **2023**, *24*, 3133–3144. [CrossRef]

14. Carpenter, K.; Tabei, A. On Residual Stress Development, Prevention, and Compensation in Metal Additive Manufacturing. *Materials* **2020**, *13*, 255. [CrossRef] [PubMed]

15. Mirsayar, M. A generalized criterion for fatigue crack growth in additively manufactured materials—Build orientation and geometry effects. *Int. J. Fatigue* **2021**, *145*, 106099. [CrossRef]

16. Kok, Y.; Tan, X.; Wang, P.; Nai, M.; Loh, N.; Liu, E.; Tor, S. Anisotropy and heterogeneity of microstructure and mechanical properties in metal additive manufacturing: A critical review. *Mater. Des.* **2018**, *139*, 565–586. [CrossRef]

17. Li, C.; Liu, Z.Y.; Fang, X.Y.; Guo, Y.B. Residual Stress in Metal Additive Manufacturing. In Proceedings of the 4th CIRP Conference on Surface Integrity (CSI), Tianjin, China, 11–13 July 2018; pp. 348–353.

18. Safarian, A.; Subaşi, M.; Karataş, C. The effect of sintering parameters on diffusion bonding of 316L stainless steel in inserted metal injection molding. *Int. J. Adv. Manuf. Technol.* **2017**, *89*, 2165–2173. [CrossRef]

19. Choi, J.-P.; Lee, G.-Y.; Song, J.-I.; Lee, W.-S.; Lee, J.-S. Sintering behavior of 316L stainless steel micro–nanopowder compact fabricated by powder injection molding. *Powder Technol.* **2015**, *279*, 196–202. [CrossRef]

20. Thompson, Y.; Gonzalez-Gutierrez, J.; Kukla, C.; Felfer, P. Fused filament fabrication, debinding and sintering as a low cost additive manufacturing method of 316L stainless steel. *Addit. Manuf.* **2019**, *30*, 8. [CrossRef]

21. Sadaf, M.; Bragaglia, M.; Nanni, F. A simple route for additive manufacturing of 316L stainless steel via Fused Filament Fabrication. *J. Manuf. Process.* **2021**, *67*, 141–150. [CrossRef]

22. Liu, B.; Wang, Y.; Lin, Z.; Zhang, T. Creating metal parts by Fused Deposition Modeling and Sintering. *Mater. Lett.* **2020**, *263*, 4. [CrossRef]

23. Singh, G.; Missiaen, J.-M.; Bouvard, D.; Chaix, J.-M. Copper extrusion 3D printing using metal injection moulding feedstock: Analysis of process parameters for green density and surface roughness optimization. *Addit. Manuf.* **2020**, *38*, 15. [CrossRef]

24. Singh, G.; Missiaen, J.-M.; Bouvard, D.; Chaix, J.-M. Additive manufacturing of 17–4 PH steel using metal injection molding feedstock: Analysis of 3D extrusion printing, debinding and sintering. *Addit. Manuf.* **2021**, *47*, 102287. [CrossRef]

25. Kluczyński, J.; Jasik, K.; Łuszczek, J.; Sarzyński, B.; Grzelak, K.; Dražan, T.; Joska, Z.; Szachogłuchowicz, I.; Płatek, P.; Małek, M. A Comparative Investigation of Properties of Metallic Parts Additively Manufactured through MEX and PBF-LB/M Technologies. *Materials* **2023**, *16*, 5200. [CrossRef] [PubMed]

26. Santamaria, R.; Wang, K.; Salasi, M.; Iannuzzi, M.; Mendoza, M.Y.; Quadir, Z. Stress Corrosion Cracking of 316L Stainless Steel Additively Manufactured with Sinter-Based Material Extrusion. *Materials* **2023**, *16*, 4006. [CrossRef] [PubMed]

27. Sargini, M.I.M.; Masood, S.H.; Palanisamy, S.; Jayamani, E.; Kapoor, A. Additive manufacturing of an automotive brake pedal by metal fused deposition modelling. In Proceedings of the 2nd International Conference on Aspects of Materials Science and Engineering (ICAMSE), Chandigarh, India, 5–6 March 2021; pp. 4601–4605.

28. Gonzalez-Gutierrez, J.; Arbeiter, F.; Schlauf, T.; Kukla, C.; Holzer, C. Tensile properties of sintered 17-4PH stainless steel fabricated by material extrusion additive manufacturing. *Mater. Lett.* **2019**, *248*, 165–168. [CrossRef]

29. Godec, D.; Cano, S.; Holzer, C.; Gonzalez-Gutierrez, J. Optimization of the 3D Printing Parameters for Tensile Properties of Specimens Produced by Fused Filament Fabrication of 17-4PH Stainless Steel. *Materials* **2020**, *13*, 774. [CrossRef]

30. Ecker, J.V.; Dobrezberger, K.; Gonzalez-Gutierrez, J.; Spoerk, M.; Gierl-Mayer, C.; Danninger, H. Additive Manufacturing of Steel and Copper Using Fused Layer Modelling: Material and Process Development. *Powder Met. Prog.* **2019**, *19*, 63–81. [CrossRef]

31. Ren, L.; Zhou, X.; Song, Z.; Zhao, C.; Liu, Q.; Xue, J.; Li, X. Process Parameter Optimization of Extrusion-Based 3D Metal Printing Utilizing PW–LDPE–SA Binder System. *Materials* **2017**, *10*, 305. [CrossRef]

32. Lengauer, W.; Duretek, I.; Fürst, M.; Schwarz, V.; Gonzalez-Gutierrez, J.; Schuschnigg, S.; Kukla, C.; Kitzmantel, M.; Neubauer, E.; Lieberwirth, C.; et al. Fabrication and properties of extrusion-based 3D-printed hardmetal and cermet components. *Int. J. Refract. Met. Hard Mater.* **2019**, *82*, 141–149. [CrossRef]

33. Singh, P.; Balla, V.K.; Tofangchi, A.; Atre, S.V.; Kate, K.H. Printability studies of Ti-6Al-4V by metal fused filament fabrication (MF3). *Int. J. Refract. Met. Hard Mater.* **2020**, *91*, 105249. [CrossRef]

34. Eickhoff, R.; Antusch, S.; Nötzel, D.; Hanemann, T. New Partially Water-Soluble Feedstocks for Additive Manufacturing of Ti6Al4V Parts by Material Extrusion. *Materials* **2023**, *16*, 3162. [CrossRef] [PubMed]

35. Bose, A.; Reidy, J.P.; Tuncer, N.; Jorgensen, L. Processing of tungsten heavy alloy by extrusion-based additive manufacturing. *Int. J. Refract. Met. Hard Mater.* **2023**, *110*, 11. [CrossRef]

36. Li, S.; Deng, H.; Lan, X.; He, B.; Li, X.; Wang, Z. Developing cost-effective indirect manufacturing of H13 steel from extrusion-printing to post-processing. *Addit. Manuf.* **2023**, *62*, 14. [CrossRef]

37. Palmero, E.M.; Casaleiz, D.; de Vicente, J.; Hernández-Vicen, J.; López-Vidal, S.; Ramiro, E.; Bollero, A. Composites based on metallic particles and tuned filling factor for 3D-printing by Fused Deposition Modeling. *Compos. Part A Appl. Sci. Manuf.* **2019**, *124*, 105497. [CrossRef]

38. Vetter, J.; Huber, F.; Wachter, S.; Körner, C.; Schmidt, M. Development of a Material Extrusion Additive Manufacturing Process of 1.2083 steel comprising FFF Printing, Solvent and Thermal Debinding and Sintering. *Procedia CIRP* **2022**, *113*, 341–346. [CrossRef]

39. Cano, S.; Gonzalez-Gutierrez, J.; Sapkota, J.; Spoerk, M.; Arbeiter, F.; Schuschnigg, S.; Holzer, C.; Kukla, C. Additive manufacturing of zirconia parts by fused filament fabrication and solvent debinding: Selection of binder formulation. *Addit. Manuf.* **2019**, *26*, 117–128. [CrossRef]

40. Kukla, C.; Cano, S.; Kaylani, D.; Schuschnigg, S.; Holzer, C.; Gonzalez-Gutierrez, J. Debinding behaviour of feedstock for material extrusion additive manufacturing of zirconia. *Powder Met.* **2019**, *62*, 196–204. [CrossRef]

41. Tapoglou, N.; Clulow, J.; Patterson, A.; Curtis, D. Characterisation of mechanical properties of 15-5PH stainless steel manufactured through direct energy deposition. *CIRP J. Manuf. Sci. Technol.* **2022**, *38*, 172–185. [CrossRef]

42. Jiang, J.; Zhang, X.; Ma, F.; Dong, S.; Yang, W.; Wu, M.; Chang, G.; Xu, L.; Zhang, C.; Luo, F. Novel phenomena of graphene secondary dispersion and phase transformation in selective laser melting of 15-5PH/graphene composites. *Addit. Manuf.* **2021**, *47*, 102207. [CrossRef]

43. Nong, X.; Zhou, X. Effect of scanning strategy on the microstructure, texture, and mechanical properties of 15-5PH stainless steel processed by selective laser melting. *Mater. Charact.* **2021**, *174*, 111012. [CrossRef]

44. Chen, W.; Xu, L.; Hao, K.; Zhang, Y.; Zhao, L.; Han, Y.; Liu, Z.; Cai, H. Effect of heat treatment on microstructure and performances of additively manufactured 15-5PH stainless steel. *Opt. Laser Technol.* **2023**, *157*, 11. [CrossRef]

45. Wagner, M.A.; Hadian, A.; Sebastian, T.; Clemens, F.; Schweizer, T.; Rodriguez-Arbaizar, M.; Carreño-Morelli, E.; Spolenak, R. Fused filament fabrication of stainless steel structures—from binder development to sintered properties. *Addit. Manuf.* **2022**, *49*, 102472. [CrossRef]

46. Riaz, A.; Töllner, P.; Ahrend, A.; Springer, A.; Milkereit, B.; Seitz, H. Optimization of composite extrusion modeling process parameters for 3D printing of low-alloy steel AISI 8740 using metal injection moulding feedstock. *Mater. Des.* **2022**, *219*, 110814. [CrossRef]

47. Ji, C.; Loh, N.; Khor, K.; Tor, S. Sintering study of 316L stainless steel metal injection molding parts using Taguchi method: Final density. *Mater. Sci. Eng. A* **2001**, *311*, 74–82. [CrossRef]

Article

In Situ SEM, TEM, EBSD Characterization of Nucleation and Early Growth of Pure Fe/Pure Al Intermetallic Compounds

Xiaojun Zhang, Kunyuan Gao *, Zhen Wang, Xiuhua Hu, Jianzhu Wang and Zuoren Nie

Faculty of Materials and Manufacturing, Beijing University of Technology, Beijing 100124, China; tigerkin1210@163.com (X.Z.); xifengmochou@outlook.com (J.W.); zrnie@bjut.edu.cn (Z.N.)
* Correspondence: gaokunyuan@bjut.edu.cn

Abstract: The nucleation and growth processes of pure Fe/pure Al intermetallic compounds (IMCs) during heat treatment at 380 °C and 520 °C were observed through in situ scanning electron microscopy (SEM). The size of the IMCs were statistically analyzed using image analysis software. The types and distribution of IMCs were characterized using transmission electron microscopy (TEM) and electron backscattering diffraction (EBSD). The results showed that: at 380 °C, the primary phase of the Fe/Al composite intermetallic compounds was Fe_4Al_{13}, formed on the Fe side and habituated with Fe. The IMC was completely transformed from the initial Fe_4Al_{13} to the most stable Fe_2Al_5, and the Fe_2Al_5 was the habitus with Fe during the process of holding at 380 °C for 15 min to 60 min. At 380 °C, the initial growth rate of the IMC was controlled by reaction, and the growth rate of the thickness and horizontal dimensions was basically the same as 0.02–0.17 μm/min. When the IMC layer thickness reached 4.5 μm, the growth rate of the thickness changed from reaction control to diffusion control and decreased to 0.007 μm/min. After heat treatment at 520 °C (≤20 min), the growth of IMC was still controlled by the reaction, the horizontal growth rate was 0.53 μm/min, the thickness growth rate was 0.23 μm/min, and the main phase of the IMC was the Fe_2Al_5 phase at 520 °C/20 min.

Keywords: pure Fe/pure Al composites; intermetallic compounds; in situ heat treatments; habitus; growth rate

check for updates

Citation: Zhang, X.; Gao, K.; Wang, Z.; Hu, X.; Wang, J.; Nie, Z. In Situ SEM, TEM, EBSD Characterization of Nucleation and Early Growth of Pure Fe/Pure Al Intermetallic Compounds. *Materials* **2023**, *16*, 6022. https://doi.org/10.3390/ma16176022

Academic Editor: Pavel Novák

Received: 12 July 2023
Revised: 16 August 2023
Accepted: 21 August 2023
Published: 1 September 2023

1. Introduction

Fe/Al composites have the advantages of the excellent mechanical properties of Fe and corrosion resistance, good thermal conductivity, and the low density of Al, so they are more and more widely used in vehicles, ships, power stations, household appliances, and in other fields [1–4]. However, in the follow-up heat treatment process, the interface of the Fe/Al composites can very easily produce intermetallic phases, which are brittle and can reduce the bonding strength of the interface. Some researchers believe that, in IMCs, a thickness less than 10 μm has no adverse effect on joint strength and may even improve the quality of the joint. However, when the thickness exceeds 10 μm, the joint strength significantly decreases [5–7]. Therefore, in the past few decades, in order to control the growth of IMCs at the interface of Fe/Al composites, studies have found that adding elements such as Si and Zn to Al could delay the formation time of IMC at the interface [8–10]. Among them, 0.8–1.5% Si element had the best inhibitory effect on IMC formation [11–13]; it was used to control Fe/Al intermetallics, inhibit the production of brittle intermetallics [14], and change the distribution and morphology of intermetallic compounds at the interface (such as from continuous distribution to intermittent distribution or from lamellar to spherical distribution) [15]. The interfacial bonding properties of the Fe/Al composites were changed by means of controlling the intermetallic layer thickness below 10 μm [16] and reducing the grain size of the IMC layer [17]. Most of these studies focused on microalloying and IMC growth kinetics. Little attention was paid to the early stages of nucleation and growth of IMCs in Fe/Al systems and their evolution with the increase in the heat treatment temperature or reaction time. However, without a deeper understanding of the entire growth kinetics of these phases, such research would still be an expensive trial-and-error process.

In recent years, in situ experiments have become important tools for elucidating the entire phase transition process at moving interfaces. Szczepaniak [18] used a new type of in situ heating transmission electron microscope for the first time, and the formation process of the Fe_xAl_y phase at the weld interface of a friction stir welding specimen was characterized in real time. The results showed that acicular Fe_4Al_{13} was the first stable phase formed in the annealed state. Sapanathan [19] characterized the nucleation and early growth stages of Fe/Al intermetallics at 596 °C using an in situ heating device in a special scanning electron microscope with electron backscatter diffraction; the results showed that the Fe_4Al_{13} phase nucleates first, before the Fe_2Al_5 phase diffusion-controlled growth. Barbora Křivská [20] utilized in situ TEM to investigate the formation of Fe_2Al_5 at the interface through isothermal annealing above 500 °C. The growth kinetics followed the typical parabolic trend of diffusion-controlled phase transformation. The brittle intermetallic phase that formed reduced the bonding strength between the steel and aluminum. Kai Zhang [21] conducted in situ observations of the melting and solidification process of an Fe/Al/Ta eutectic alloy using high-temperature confocal scanning laser microscopy. The results showed that when the temperature was below 1600 °C, no other types of phase transformations were observed in the Fe/Al/Ta eutectic alloy. After solidification, the strengthening phase exhibited a certain orientation, and the microstructure at the center of the eutectic cell was more regular compared with the microstructure at the grain boundaries. Junsheng Wang [22] conducted a study of the in situ synchrotron radiation imaging of the formation of iron-rich intermetallics during the solidification process of an Al-7.5Si-3.5Cu-0.8Fe (wt.%) alloy. It was found that the nucleation of iron-rich β-intermetallic compounds occurred between 550 and 570 °C. Initially, they grew with an instantaneous tip velocity of 100 μm/s, which then slowed down to 10 μm/s at the end of the growth.

Due to the rapid growth kinetics of IMCs in Fe/Al binary systems, the early stages of nucleation and IMC growth cannot be captured by non-in situ analysis. Therefore, this study employed in situ heating SEM observations at 380 °C and 520 °C to monitor the microstructural and morphological changes in a Fe/Al system, aiming to investigate the nucleation, early growth kinetics, and phase transition of IMC in the initial stages, and quantitative analysis was conducted to study the growth process of the multiple nucleation points of the IMC. The ultimate goal was to explore the growth mechanism of IMC.

2. Materials and Methods

The materials were 2 mm annealed pure Al plate (99.99 wt.%) and 3 mm annealed pure Fe plate (99.9 wt.%). Before the rolling composite, the Al plate and Fe plate were pickled, the composite surface was polished with a steel brush, and the rolling deformation was 40%; after the final rolling, the thickness of the Fe/Al composite was 3 mm, the Al layer was 1 mm, and the Fe layer was 2 mm. The cold-rolled composite was carried out at 25 °C; the Fe layer would not oxidize, and the Al layer material would oxidize and produced Al_2O_3 after grinding. The thickness of the oxide layer was 2–3 nm, and the oxide layer was broken during the rolling process, which did not adversely affect the rolled composite interface. The wire-cut sample size was $460 \times 30 \times 1$ mm^3 (length $\times$ width $\times$ height), and a SEM (Gemini SEM 300 (Zeiss, Oberkochen, Germany)) equipped with a heating table (MINI-HT1200-SE, as shown in Figure 1) was used to heat the sample in situ at 380 °C and 520 °C. The temperature was continuously measured with the thermocouple in contact with the sample, the temperature control accuracy was ±2 °C, and the working voltage of the SEM was 18kV. The microstructure evolution during the heating process was continuously recorded using screen recording software (EVCapture v4.0.2).

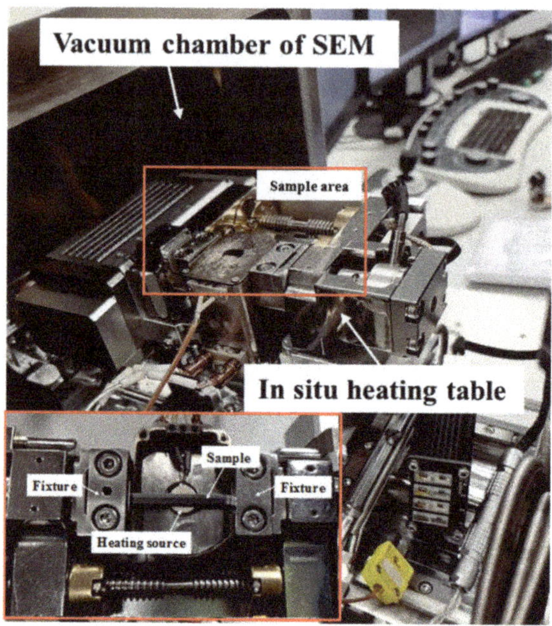

Figure 1. In situ heating unit.

In our early non-in situ experiments, it was found that after a 370 °C/1 h heat treatment of pure Fe/pure Al composites, the thickness of IMCs was approximately 10 nm, making it challenging to observe the nucleation and growth process using SEM. These results will be presented in another article. However, after a 380 °C/1 h heat treatment, the IMCs thickness increased to approximately 3–5 μm, which met the size requirements for in situ SEM observation during the heating process. Moreover, at 380 °C, the growth rate of the IMCs layer was slow, which was highly favorable for clear observation of the details of IMCs nucleation and subsequent growth processes. Therefore, 380 °C was selected as the minimum temperature for in situ heating experiments, and 520 °C was the heat treatment temperature of materials used for air cooling in power stations [2].

After the heating stage, because the thickness of intermetallic compound layers fluctuated greatly, in order to better study the growth law of intermetallic compound layers, the maximum values in 20 fields of view were counted in a sample, and the average values were used to characterize the average thickness of intermetallic compound layers, then the intermetallic layer size was evaluated by image analysis system (image pro plus 6.0), FIB sample preparation (Helios G4 PFIB CXe (Thermo Fisher Scientific, Waltham, MA, USA)) was used for TEM (FEITalosF200X-G2 (Lincoln, NE, USA)) microscopic analysis, and after fine polishing and vibration polishing, EBSD analysis (FEI QUANTA FEG650 (Thermo Fisher Scientific, Waltham, MA, USA)) was carried out, as shown in Table 1.

Table 1. Samples in materials and methods.

Sample	Materials	Temperature/°C	Holding Time/min	Characterization Methods
1#		380	20	FIB + TEM
2#	pure Fe/pure Al	380	60	FIB + TEM, Size statistics
3#		520	20	EBSD, Size statistics

3. Results and Discussion

3.1. Nucleation and Characterization of Fe/Al Intermetallic Compounds

Figure 2 shows the SEM and TEM characterization of Fe/Al composites intermetallic compounds during in situ heating at 380 °C. Figure 2a shows the interface morphology of Fe/Al composites without heating, Figure 2b shows the enlarged interface morphology of the green box in Figure 2a after heat treatment at 380 °C/20 min. It could be seen from Figure 2b that IMCs of about 100 nm were formed at the interface after heat treatment. The TEM analysis of IMCs was shown in Figure 2c,d. As showed in Figure 2d, the red markings represent the diffraction pattern of Fe, while the green markings represent the diffraction pattern of Fe_4Al_{13}, and it can be seen that the IMCs were Fe_4Al_{13} phase, the (2 0 0) plane of Fe was parallel to the (6 0 1) plane of Fe_4Al_{13}, and the axial direction of Fe [0 0 1] was parallel to the axial direction of Fe_4Al_{13} [1,3–6], description of the initial Fe_4Al_{13} and Fe habitus. Therefore, the primary phase of Fe/Al composites intermetallic compounds was the Fe_4Al_{13}.

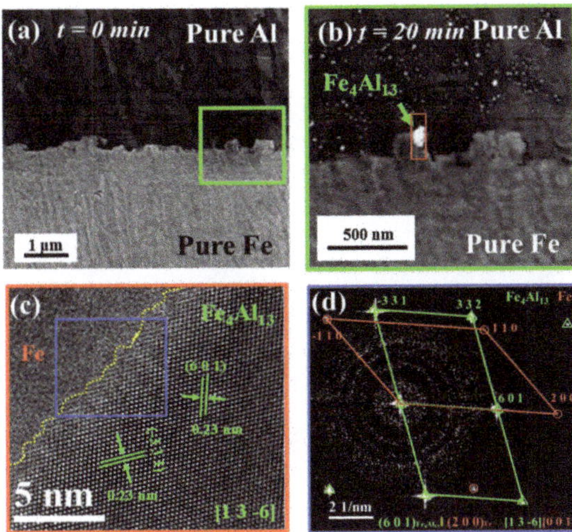

Figure 2. Fe/Al in situ heat treatment with SEM and TEM at 380 °C/20 min. (**a**) In situ SEM of Fe/Al composites at 380 °C/0 min. (**b**) In situ SEM of Fe/Al composites at 380 °C/20 min. (**c,d**) TEM of Fe/Al composites at 380 °C/20 min.

3.2. In Situ Observation and Characterization of Early Growth of IMCs at 380 °C

Figure 3a–i show the nucleation and growth of IMCs during heat treatment at 380 °C, and as can be seen from the picture, IMCs (about 400 nm) began to nucleate at the A position was on the Fe side. With the extension of holding time, three new nucleation points, B, C and D, were formed at the interface, and the thickness was between 100–400 nm. The IMCs at four points had been integrated in the horizontal direction, especially the IMCs at points B and D, which were difficult to distinguish under SEM at 380 °C/60 min. In order to facilitate subsequent size statistics, the horizontal dimension of the IMCs at points B and D was combined into $B_{width} + D_{width}$.

The maximum size of IMCs in the horizontal and thickness direction at points A, B, C and D in Figure 3 were statistically analyzed using image pro, and the growth degree of IMCs at each point was quantitatively measured, as shown in Figure 4. The initial growth rate of IMCs at the four position points was basically consistent in the thickness and horizontal direction. The initial growth rate of IMCs at point A was 0.17 μm/min, point B and point C were all 0.07 μm/min, and point D was 0.02 μm/min. It was worth noting that after 50 min at point A, the growth rate of IMCs in the thickness direction was greatly reduced to 0.007 μm/min, but the growth rate in the horizontal direction had little change.

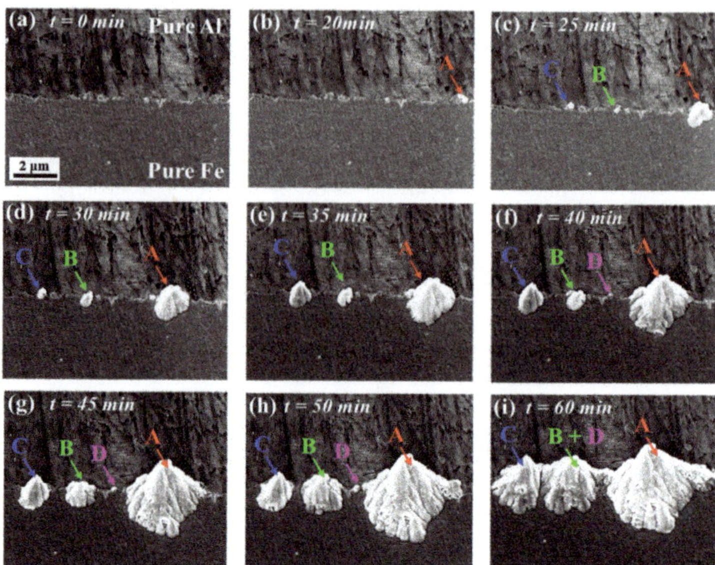

Figure 3. Sequence of IMCs nucleation and growth obtained from in situ SEM observations at 380 °C (**a**) t = 0 min; (**b**) t = 20 min, the red arrow indicates where the interface IMCs began to nucleate and marked it as A; (**c**) t = 25 min, two nucleation points B and C were added at the interface; (**d**) t = 30 min; (**e**) t = 35 min; (**f**) t = 40 min, a D-nucleate point was added at the interface; (**g**) t = 45 min; (**h**) t = 50 min; (**i**) t = 60 min, the IMCs of point B and D grew into one body, and then the IMCs of two points B and D were analyzed as B + D.

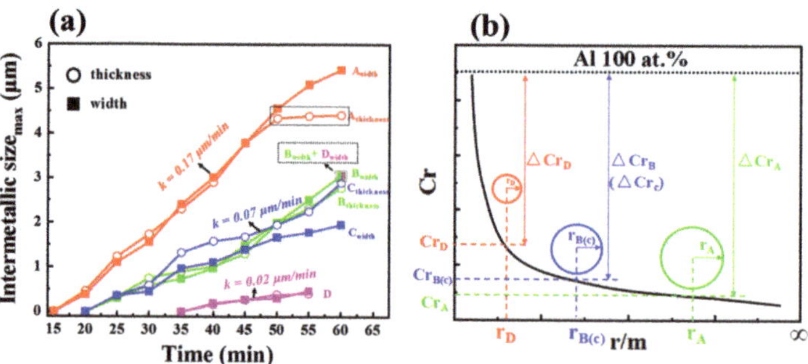

Figure 4. (**a**) Evolution of the maximum size of IMCs phase in horizontal direction and thickness direction over time at points A, B, C and D (in Figure 3); (**b**) the Cr with changing of r (considering the Gibbs–Thomson effect).

At the initial stage of IMCs growth, the change in IMCs thickness was controlled by reaction and had a linear relationship with time. In this study, the horizontal size and thickness growth rate were basically the same, which was also linear with time. Therefore, in the initial stage of IMCs growth, the horizontal size of IMCs was also controlled by the reaction. When the thickness of IMCs reached 4.5 μm, the growth rate of thickness decreased greatly, which was due to the limitation of element diffusion, and the growth mode changed from reaction control to diffusion control, so, at 380 °C, the critical thickness of the transition from the reaction-controlled to diffusion-controlled growth rate was 4.5 μm.

When the second phase particles were very small, they had a high surface-to-volume ratio and a small curvature radius. In this case, the influence of surface tension on solubility must be considered, which is known as the Gibbs–Thomson effect [23]. Indeed, the corrected solubility limit C_r of B atoms in α matrix in equilibrium with β phase occurring as spherical particles of radius r is often given as a function of r [24]:

$$C_r^\alpha = C_\infty^\alpha exp\left(\frac{2\gamma V_m}{rRT}\right) \tag{1}$$

where C_r is the solute concentration at the interfacial region in the matrix near the second phase. r is the radius of the second phase particles. T is the temperature, γ is the surface energy, R is the molar gas constant and V_m is the molar volume.

Considering the Gibbs–Thomson effect, it was demonstrated in Figure 4b schematically the solute concentration C_r with changing of particle radius. Because the radii of the second phase particles at four positions pointed in Figure 3 was in the order of $r_A > r_B$ (or r_C) $> r_D$, the solute concentration near the particles could be estimated based on Equation (1) as $C_{rD} > C_{rB}$ (or C_{rC}) $> C_{rA}$. It is clear that the solute concentration in the Al matrix was 100%, thereafter the solute concentration difference between C_r and Al matrix was $\Delta C_r A > \Delta C_{rB}$ (or ΔC_{rC}) $> \Delta C_{rD}$. Correspondingly, the chemical driving force for the second phase was greatest at point A and least at point D; therefore, the growth rates of the second phase at the four points were evaluated to be $k_A > k_B$ (or k_C) $> k_D$.

It should be noted that the surface of the sample heated in situ observed in this study was an open free surface, which was easier to nucleate and grow than the inside of the sample, and its morphology was irregular polygon-like. In the subsequent TEM and EBSD characterization, it was found that the size of IMCs on the non-free surface inside the interface was significantly smaller than that on the free surface, and its morphology was flat and its thickness was evenly distributed.

Figure 5 shows the TEM analysis of the interface IMCs after heat treatment at 380 °C/60 min, and Figure 5a shows the interface morphology of the sample after fine grinding and polishing. The thickness of the IMCs layer was about 500 nm, and the orange box represented the FIB sample preparation area. Figure 5b shows the sample prepared by FIB, and it can be seen that there are two obvious pore defects in the IMCs layer. IMCs was closely bound to the contact surface of Fe layer and Al layer, without obvious gaps, cracks, and other defects. Figure 5c shows the TEM bright field diagram, and it can be seen that cracks and other defects appeared at the contact surface between the IMCs and Fe layer, as well as between the IMCs and Al layer, which should be generated during the thinning process of the FIB sample preparation and not the defects of the sample itself. The diffraction analysis of IMCs was shown in Figure 5f, the IMCs were polycrystalline Fe_2Al_5. We performed high-resolution analysis on the interface between IMCs layer and Fe layer (green color box in Figure 5c) and the interface between IMCs layer and Al layer (red color box in Figure 5c), respectively, and the results showed that the (2 1 1) and (2 2 0) planes of the Fe_2Al_5 phase were parallel to the (1 0 $\bar{1}$) and (1 1 0) planes of Fe, respectively, and Fe_2Al_5 was habituated with Fe, but no parallel plane existed between Fe_2Al_5 and Al.

According to reports [5–7], when the thickness of IMCs was less than 4 μm, and no cracks or other defects on the contact surface of the IMCs and Fe layer, IMCs and Al layer, the bonding strength of IMCs relative to Fe/Al composites interface had no adverse effects, in this study, at 380 °C/60 min, the IMCs were about 500 nm, and there were no cracks and other defects on the contact surface, so the IMCs had no adverse effect on the bonding strength of the Fe/Al composites interface. The IMCs of Fe/Al composites were completely transformed from the initial Fe_4Al_{13} to the most stable Fe_2Al_5 during heat treatment at 380 °C for 15 min to 60 min. Both the Fe_4Al_{13} and Fe_2Al_5 phases were habituated with Fe, indicating that Fe_4Al_{13} and Fe_2Al_5 phases had a closer orientation to Fe and were generated from the Fe side.

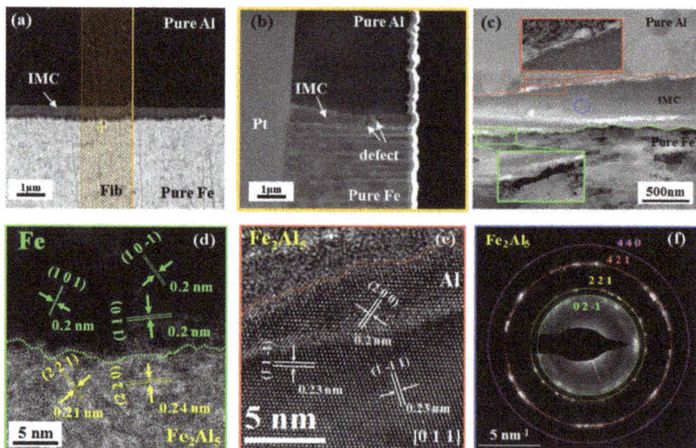

Figure 5. TEM analysis of IMCs after heat treatment at 380 °C/60 min. (**a**) SEM interface morphology of IMCs after fine grinding and polishing; (**b**) SEM morphology after FIB preparation; (**c**) TEM sample open field diagram; (**d**) high resolution image at the interface between IMCs layer and Fe layer; (**e**) high resolution image at the interface between IMCs layer and Al layer; (**f**) diffraction pattern of IMCs phase.

3.3. In Situ Observation and Characterization of IMCs at 520 °C

Figure 6a–i showed the nucleation and growth of IMCs during heat treatment at 520 °C; as seen from Figure 6b, when the holding time was 4 min, the IMCs began to nucleate at the interface, and the IMCs were all less than 1 μm, at the same time, the horizontal size of IMCs on the left side of point A was L_{width}. When the holding time was extended to 6 min, more nucleation occurred at the interface, and obvious crack defects were formed on the Al side near the IMCs layer, as indicated by the yellow arrow in Figure 6c–i. With the extension of holding time, the size of IMCs gradually increased along the thickness and horizontal direction, and the cracks and defects concentrated in the Al side near the IMCs increased. The average thickness size and L_{width} of the interface IMCs in the whole process were statistically analyzed, and the results were shown in Figure 7.

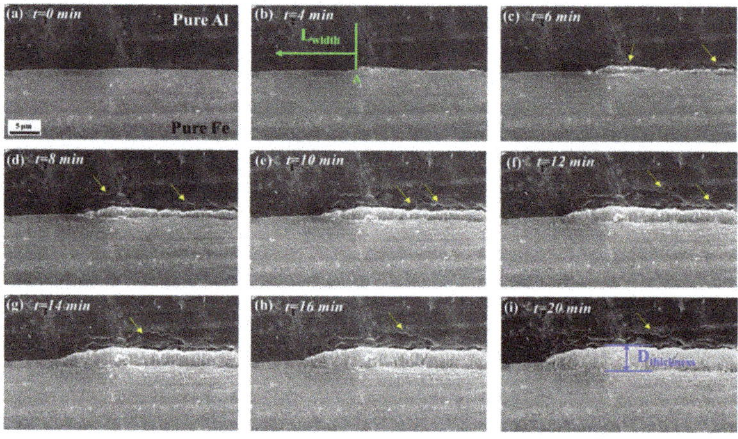

Figure 6. Sequence of IMCs nucleation and growth obtained from in situ SEM observations at 520 °C (**a**) t = 0 min; (**b**) t = 4 min, the horizontal dimension of the IMCs layer to the left of point A was L_{width}; (**c**) t = 6 min; (**d**) t = 8 min; (**e**) t = 10 min; (**f**) t = 12 min; (**g**) t = 14 min; (**h**) t = 16 min; (**i**) t = 20 min.

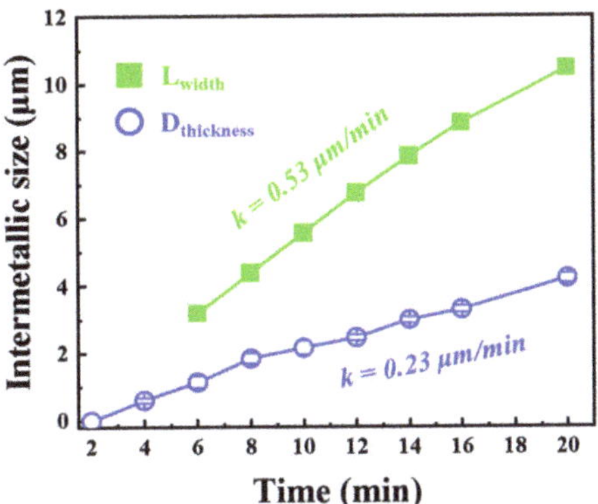

Figure 7. Evolution of IMCs phase average thickness and maximum L_{width} size over time.

It could be seen from Figure 7 that the average thickness growth rate of IMCs was 0.23 μm/min, and the growth rate of L_{width} was 0.53 μm/min, which was 2.3 times larger than that of the thickness. Compared with 380 °C, at 520 °C heat treatment, the growth of IMCs phase size was still controlled by the reaction, but its growth rate in the horizontal direction was significantly higher than that in the thickness direction.

There were obvious crack defects on the Al side, which might be caused by two reasons [25]: ① The Kirkendall effect during the diffusion of elements; ② The volume expansion caused by IMCs generation. A single Al cell was composed of 4 Al atoms, the volume of the cell was $V_{Al} = 66.3\ \text{Å}^3$, a single Fe cell consisted of 2 Fe atoms, and the cell volume $V_{Fe} = 23.6\ \text{Å}^3$, a single Fe_2Al_5 cell consisted of 10.8 Al atoms and 4 Fe atoms, and the cell volume $V_{Fe2Al5} = 206.5\ \text{Å}^3$. Because the difference between V_{Fe2Al5} and the consumed V was less than 10%, the volume change in the generated Fe_2Al_5 phase was not the main reason for the crack defects.

At 520 °C, $D_{Al}\ (Fe_2Al_5) = 8.1 \times 10^{-6}\ \text{cm}^2/\text{s} > D_{Fe}\ (Fe_2Al_5) = 3.6 \times 10^{-12}\ \text{cm}^2/\text{s}$ [26], the number of atoms of Al atom arriving at the Fe interface through Fe_2Al_5 was six orders of magnitude higher than that of Fe atom arriving at the Al interface through Fe_2Al_5 phase, so the number of diffused Fe atoms was negligible, and the diffusion of Al in Fe_2Al_5 was dominant. In the subsequent process of this experiment (at 520 °C/60 min), the crack defects would be transformed into separated cracks, as the sample became more continuous during cooling (the results of this part of the experiment will be shown in another article), and it could be inferred that the crack defects were mainly caused by the Al side of Kirkendall.

Figure 8 shows the EBSD characterization of Fe/Al composites after in situ heating at 520 °C/20 min, RD surface with fine grinding and vibration polishing. It can be seen that the main constituent phase of IMCs was Fe_2Al_5, and only a small amount of Fe_4Al_{13} was distributed in the middle of Fe_2Al_5. It was confirmed that Fe_2Al_5 was the main phase and stable phase of IMCs in Fe/Al composites.

During the heat treatment process, the Fe_4Al_{13} was the primary phase that formed at the Fe/Al interface initially; subsequently, the Fe_4Al_{13} underwent rapid reaction with the Fe, transforming into the Fe_2Al_5; therefore, the Fe_2Al_5 rapidly grew and became the dominant phase of IMCs. Moreover, the diffusion rate of Al within the Fe_2Al_5 was significantly higher than that of Fe within the Fe_2Al_5 [26], the new Fe_4Al_{13} forms at the Al/Fe_2Al_5 interface [27], the coexistence of two IMCs, with the Fe_2Al_5 as the dominant one, had been observed. In Figure 8b, an inverse pole figure (IPF) map was presented, showing that the Al matrix was in a recrystallized state, while the Fe matrix was still in a deformed state. Concerning Fe_2Al_5, at 520 °C, the reaction might locate at the proper value region of driving force and diffusion rate, i.e., the nose of the reaction, which could result in smaller grain sizes.

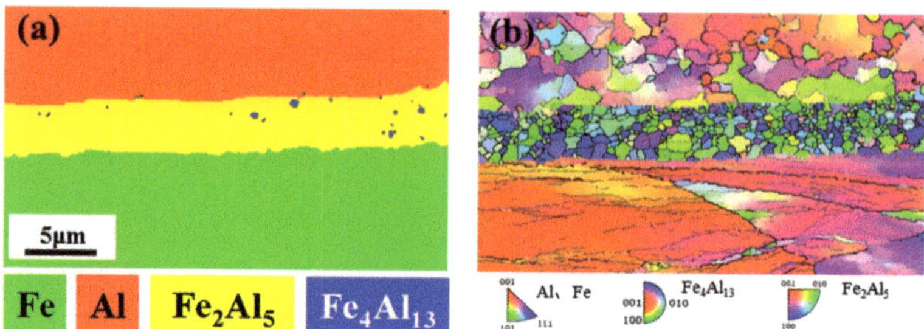

Figure 8. (**a**,**b**) EBSD characterization of IMCs at 520 °C/20 min.

4. Conclusions

(1)   At 380 °C, the primary phase of Fe/Al composite intermetallic compounds was Fe_4Al_{13} formed on the Fe side and habituated with Fe. The IMCs changed from the initial Fe_4Al_{13} to the most stable Fe_2Al_5 when the heat treatment was extended from 15 min to 60 min, and the Fe_2Al_5 phase was habitus with Fe.

(2)   During heat treatment at 380 °C, the initial growth rate of Fe/Al composite intermetallic compounds was controlled by reaction. The initial growth rate of thickness and horizontal dimensions of IMCs was basically the same, ranging from 0.02–0.17 μm/min. When the thickness reached 4.5 μm, the growth rate of the thickness changed from reaction control to diffusion control, the critical thickness of the transformation was 4.5 μm, and the growth rate decreased to 0.007 μm/min.

(3)   After heat treatment at 520 °C ($\leq$20 min), the growth of IMCs was still controlled by the reaction, the horizontal growth rate was 0.53 μm/min, the thickness growth rate was 0.23 μm/min, and the main component of IMCs was Fe_2Al_5 at 520 °C/20 min.

Author Contributions: X.Z.: data curation, investigation, writing—original draft. Z.W.: data curation, X.H.: investigation. J.W.: resources. K.G.: writing—review and editing. Z.N.: supervision and editing. All authors have read and agreed to the published version of the manuscript.

Funding: This research was funded by National Key Research and Development Program of China (2022YFB3705802, 2021YFB3704201 and 2021YFB3704202), Jiangsu Province Program for Commercialization of Scientific and Technological Achievements (BA2022029) and Program on Jiangsu Key Laboratory for Clad Materials (BM2014006).

Institutional Review Board Statement: Not applicable.

Informed Consent Statement: Not applicable.

Data Availability Statement: This study did not report any data.

Conflicts of Interest: The authors declare no conflict of interest.

References

1.   Yu, G.; Chen, S.; Li, S.; Huang, J.; Yang, J.; Zhao, Z.; Huang, W.; Chen, S. Microstructures and mechanical property of 5052 aluminum alloy/Q235 steel butt joint achieved by laser beam joining with Sn-Zn filler wire. *Opt. Laser Technol.* **2021**, *139*, 106996. [CrossRef]

2.   Chen, X.; Li, L.; Zhou, D.J. Formation and bonding properties of Al(4A60)-steel(08Al) clad strip intermetallic compound. *Zhongguo Youse Jinshu Xuebao/Chin. J. Nonferrous Met.* **2015**, *25*, 1176–1184.

3.   Sasaki, T.; Yakou, T. Features of intermetallic compounds in aluminized steels formed using aluminum foil. *Surf. Coat. Technol.* **2006**, *201*, 2131–2139. [CrossRef]

4.   Matysik, P.; Jozwiak, S.; Czujko, T. Characterization of Low-Symmetry Structures from Phase Equilibrium of Fe-Al System-Microstructures and Mechanical Properties. *Materials* **2015**, *8*, 914–931. [CrossRef] [PubMed]

5.   Tanaka, Y.; Kajihara, M. Morphology of Compounds Formed by Isothermal Reactive Diffusion between Solid Fe and Liquid Al. *Mater. Trans.* **2009**, *50*, 2212–2220. [CrossRef]

6.   Zhang, X.; Gao, K.; Wang, Z.; Hu, X.; Liu, H.; Nie, Z. Effect of intermetallic compounds on interfacial bonding of Al/Fe composites. *Mater. Lett.* **2023**, *333*, 133597. [CrossRef]

7. Shin, D.; Lee, J.-Y.; Heo, H.; Kang, C.-Y. Formation Procedure of Reaction Phases in Al Hot Dipping Process of Steel. *Metals* **2018**, *8*, 820. [CrossRef]
8. Zou, T.P.; Gao-Yang, Y.U.; Chen, S.H.; Huang, J.H.; Yang, J.; Zhao, Z.Y.; Rong, J.P.; Yang, J. Effect of Si content on interfacial reaction and properties between solid steel and liquid aluminum. *Trans. Nonferrous Met. Soc. China* **2021**, *31*, 2570–2584. [CrossRef]
9. Krivska, B.; Slapakova, M.; Vesely, J.; Kihoulou, M.; Fekete, K.; Minarik, P.; Kralik, R.; Grydin, O.; Stolbchenko, M.; Schaper, M. Intermetallic Phases Identification and Diffusion Simulation in Twin-Roll Cast Al-Fe Clad Sheet. *Materials* **2021**, *14*, 7771. [CrossRef]
10. Zhang, P.; Shi, H.; Tian, Y.; Yu, Z.; Wu, D. Effect of zinc on the fracture behavior of galvanized steel/6061 aluminum alloy by laser brazing. *Weld. World* **2021**, *65*, 13–22. [CrossRef]
11. Xia, H.; Zhao, X.; Tan, C.; Chen, B.; Song, X.; Li, L. Effect of Si content on the interfacial reactions in laser welded-brazed Al/steel dissimilar butted joint. *J. Mater. Process. Technol.* **2018**, *258*, 9–21. [CrossRef]
12. Maitra, T.; Gupta, S.P. Intermetallic compound formation in Fe–Al–Si ternary system: Part II. *Mater. Charact.* **2002**, *49*, 293–311. [CrossRef]
13. Dangi, B.; Brown, T.W.; Kulkarni, K.N. Effect of silicon, manganese and nickel present in iron on the intermetallic growth at iron—Aluminum alloy interface. *J. Alloys Compd.* **2018**, *769*, 777–787. [CrossRef]
14. Bouayad, A.; Gerometta, C.; Belkebir, A.; Ambari, A. Kinetic interactions between solid iron and molten aluminium. *Mater. Sci. Eng. A* **2003**, *363*, 53–61. [CrossRef]
15. Hirose, A.; Imaeda, H.; Kondo, M.; Kobayashi, K.F. Influence of Alloying Elements on Interfacial Reaction and Strength of Aluminum/Steel Dissimilar Joints for Light Weight Car Body. *Mater. Sci. Forum* **2007**, *539–543*, 3888–3893. [CrossRef]
16. Yahiro, A.; Masui, T.; Yoshida, T.; Doi, D. Development of Nonferrous Clad Plate and Sheet by Warm Rolling with Different Temperature of Materials. *Trans. Iron Steel Inst. Jpn.* **2007**, *31*, 647–654. [CrossRef]
17. Furuya, H.S.; Sato, Y.T.; Sato, Y.S.; Kokawa, H.; Tatsumi, Y. Strength Improvement Through Grain Refinement of Intermetallic Compound at Al/Fe Dissimilar Joint Interface by the Addition of Alloying Elements. *Metall. Mater. Trans. A* **2017**, *49*, 527–536. [CrossRef]
18. Szczepaniak, A. Investigation of Intermetallic Layer Formation in Dependence of Process Parameters during the Thermal Joining of Aluminium with Steel. Ph.D. Thesis, Rheinisch-Westfälischen Technischen Hochschule Aachen, Aachen, Germany, 2016; pp. 36–40.
19. Sapanathan, T.; Sabirov, I.; Xia, P.; Monclús, M.A.; Molina-Aldareguía, J.M.; Jacques, P.J.; Simar, A. High temperature in situ SEM assessment followed by ex situ AFM and EBSD investigation of the nucleation and early growth stages of Fe-Al intermetallics. *Scr. Mater.* **2021**, *200*, 113910. [CrossRef]
20. Křivská, B.; Šlapáková, M.; Minárik, P.; Fekete, K.; Králík, R.; Stolbchenko, M.; Schaper, M.; Grydin, O. In-Situ TEM observation of intermetallic phase growth in Al-steel clad sheet. *AIP Conf. Proc.* **2021**, *2411*, 030005. [CrossRef]
21. Zhang, K.; Cui, C.; Deng, L.; Wang, Y.; Wu, C.; Su, H. In-situ observations of solidification process of Fe–Al–Ta eutectic alloy. *J. Phys. Chem. Solids* **2023**, *172*, 111067. [CrossRef]
22. Wang, J.; Lee, P.; Hamilton, R.; Li, M.; Allison, J. The kinetics of Fe-rich intermetallic formation in aluminium alloys: In situ observation. *Scr. Mater.* **2009**, *60*, 516–519. [CrossRef]
23. Porter, D.A.; Easterling, K.E. *Phase Transformation in Metals and Alloys*; Chapman and Hall: London, UK, 1992; p. 514.
24. Perez, M. Gibbs-Thomson effects in phase transformations. *Scr. Mater.* **2005**, *52*, 709–712. [CrossRef]
25. Springer, H.; Kostka, A.; dos Santos, J.F.; Raabe, D. Influence of intermetallic phases and Kirkendall-porosity on the mechanical properties of joints between steel and aluminium alloys. *Mater. Sci. Eng. A* **2011**, *528*, 4630–4642. [CrossRef]
26. Eggersmann, M.; Mehrer, H. Diffusion in intermetallic phases of the Fe-Al system. *Philos. Mag. A* **2000**, *80*, 1219–1244. [CrossRef]
27. Lee, J.-M.; Kang, S.-B.; Sato, T.; Tezuka, H.; Kamio, A. Evolution of iron aluminide in Al/Fe in situ composites fabricated by plasma synthesis method. *Mater. Sci. Eng. A* **2003**, *362*, 257–263. [CrossRef]

Article

Laser Powder Bed Fusion of Inconel 718 Tools for Cold Deep Drawing Applications: Optimization of Printing and Post-Processing Parameters

Cho-Pei Jiang [1,2,*], Andi Ard Maidhah [3], Shun-Hsien Wang [4], Yuh-Ru Wang [5], Tim Pasang [6] and Maziar Ramezani [7]

[1] Department of Mechanical Engineering, National Taipei University of Technology, Taipei 10608, Taiwan
[2] High-Value Biomaterials Research and Commercialization Center, National Taipei University of Technology, Taipei 10608, Taiwan
[3] College of Mechanical and Electrical Engineering, National Taipei University of Technology, Taipei 10608, Taiwan; ardmaidhah13@gmail.com
[4] Graduate Institute of Mechatronics, National Taipei University of Technology, Taipei 10608, Taiwan
[5] Department of Materials and Mineral Resources Engineering, National Taipei University of Technology, Taipei 10608, Taiwan; yavis6509wang@gmail.com
[6] Department of Manufacturing and Mechanical Engineering Technology, Oregon Institute of Technology, 3201 Campus Drive, Klamath Falls, OR 97601, USA; tim.pasang@oit.edu
[7] Department of Mechanical Engineering, Auckland University of Technology, 55 Wellesley Street East, Auckland 1010, New Zealand; maziar.ramezani@aut.ac.nz
*   Correspondence: jcp@mail.ntut.edu.tw

Citation: Jiang, C.-P.; Maidhah, A.A.; Wang, S.-H.; Wang, Y.-R.; Pasang, T.; Ramezani, M. Laser Powder Bed Fusion of Inconel 718 Tools for Cold Deep Drawing Applications: Optimization of Printing and Post-Processing Parameters. *Materials* **2023**, *16*, 4707. https://doi.org/10.3390/ma16134707

Academic Editor: Damon Kent

Received: 18 May 2023
Revised: 19 June 2023
Accepted: 22 June 2023
Published: 29 June 2023

Abstract: Inconel 718 (IN 718) powder is used for a laser powder bed fusion (LPBF) printer, but the mechanical properties of the as-built object are not suited to cold deep drawing applications. This study uses the Taguchi method to design experimental groups to determine the effect of various factors on the mechanical properties of as-built objects produced using an LPBF printer. The optimal printing parameters are defined using the result for the factor response to produce an as-built object with the greatest ultimate tensile strength (UTS), and this is used to produce a specimen for post-processing, including heat treatment (HT) and surface finishing. The HT parameter value that gives the maximum UTS is the optimal HT parameter. The optimal printing and HT parameter values are used to manufacture a die and a punch to verify the suitability of the manufactured tool for deep drawing applications. The experimental results show that the greatest UTS is 1091.33 MPa. The optimal printing parameters include a laser power of 190 W, a scanning speed of 600 mm/s, a hatch space of 0.105 mm and a layer thickness of 40 μm, which give a UTS of 1122.88 MPa. The UTS for the post-processed specimen increases to 1511.9 MPa. The optimal parameter values for HT are heating to 720 °C and maintaining this temperature for 8 h, decreasing the temperature to 620 °C and maintaining this temperature for 8 h, and cooling to room temperature in the furnace. Surface finishing increases the hardness to HRC 55. Tools, including a punch and a die, are manufactured using these optimized parameter values. The deep drawing experiment demonstrates that the manufactured tools that are produced using these values form a round cup of Aluminum alloy 6061. The parameter values that are defined can be used to manufacture IN 718 tools with a UTS of more than 1500 MPa and a hardness of more than 50 HRC, so these tools are suited to cold deep drawing specifications.

Keywords: power bed fusion; Inconel 718; deep drawing; optimization; post-processing

1. Introduction

Additive manufacturing (AM) is a layer-stacking technology that forms material, layer by layer, into a three-dimensional physical object. AM is used to manufacture complex geometrical objects with low material waste and produces objects faster than a subtractive manufacturing method. AM is used in the biomedical, automotive, aerospace and tooling industries. Powder bed fusion (PBF), sheet lamination (SL), binder jetting (BJ), material extrusion (ME), and direct energy deposition (DED) are techniques to print metal parts. PBF is the most common method because it is financially expedient, and production time is reduced. It also produces metal objects with acceptable accuracy and with mechanical properties that are similar to those of parts that are produced using conventional manufacturing methods [1–4].

Laser powder bed fusion (LPBF) is used to print metal objects with complex geometries. LPBF uses a high-intensity laser to melt metal powder using a predefined scanning path for each layer, so layers are stacked to form a three-dimensional (3D) object. The printing parameters include laser power, scanning speed, hatch space, and layer thickness, and these are defined before the printing process is conducted [5–7]. These four parameters affect the relative density, surface roughness, and mechanical properties of an as-built object that is produced using an LPBF printer.

Inconel (IN) 718 powder is a nickel-based super-alloy that is used for LPBF printers [1,6–9]. It features high wear resistance, superior corrosion resistance, and excellent mechanical properties that remain stable at high temperatures. It is used to manufacture turbine blades for the power industry, jet turbines for the aerospace industry, and to manufacture mould and die parts for metal forming because mechanical properties are improved [10,11]. The results of several studies on the printing parameters for IN 718 are summarized in Table 1, which shows that different LPBF printers use different printing parameters. The values for printing parameters must be optimized to maximize accuracy and to optimize the good mechanical properties of an as-built object. The optimization of printing parameters by trial-and-error wastes time and money.

Table 1. Printing parameters for IN 718 for different LPBF machines.

Laser Power (W)	Scanning Speed (mm/s)	Hatch Space (μm)	Layer Thickness (μm)	Ref.
180	600	105	35	[12]
200	700	-	60	[13]
350	600	80	40	[14]
200	800	105	30	[15]
190	800	90	30	[16]
180	600	105	30	[17]
170–370	500–1200	80–120	40	[18]

The optimization of parameter values for a material for an LPBF printer requires a long research process [19,20]. The Taguchi method can be used to optimize the printing parameter values for an LPBF printer for specific materials. The time and material cost [21] for an experiment is reduced because a design of experiment replaces trial and error [22]. One study's results for the printing parameters for IN 625 for an LPBF show that the laser power most significantly affects hardness and surface roughness, followed by the scanning speed and hatch space [23]. Another study determined the effect of laser power, scan speed, and hatch space on the micro-hardness and surface roughness of printed IN 625 samples, and the results show that the optimal values for printing parameters are a laser power of 270 W, a scan speed of 800 mm/s and a hatch space of 0.08 mm, which produce a micro-hardness of 416 HV and a surface roughness of 2.82 μm [21]. Analysis of variance (ANOVA) is used in many of the Taguchi methods as a test procedure to solve the problem of optimizing parameters for an output in a process [24]. SS 316L was fabricated with LPBF using ANOVA as a test procedure, verifying the effect of printing parameters in reducing porosity formed in the fabrication process. Their study result shows a 74.9% reduction in porosity [25]. However, no studies use the Taguchi method to optimize the values for printing parameters for IN 718 or to determine the factors that affect the ultimate tensile strength of the as-built object.

AM is used for the direct printing of metal dies [26]. LPBF was also used to print die-casting inserts, and the mechanical properties of the printed inserts render them unsuited for use in the production of a die. Printed insert dies must be post-processed to improve the mechanical properties of the as-built object [27]. Cold deep drawing is one type of metal formed in the manufacturing process. This cold working process can produce metal parts at temperatures from room temperature to 30% of the melting point of the material being worked. The mechanical properties of a cold drawing die include an ultimate tensile strength (UTS) of more than 1500 MPa and a hardness of more than 50 HRC [28,29].

Another study printed a die using H13 for hot extrusion using an LPBF printer, and the experimental results show that there are high residual stresses in the as-built H13 die, so it is unsuitable for direct extrusion because high residual stress leads to cracking easily when the material deforms in the extrusion die [30]. This shows that the mechanical properties of as-built metal die that are produced using an LPBF printer must be measured to determine their suitability for die applications. Post-processing improves the mechanical properties of the as-built object.

Heat treatment (HT) is used in the manufacturing industry to improve the mechanical properties of metals. IN 718 can undergo heat treatment using precipitate hardening. Generally, precipitate hardening has two steps: the first is solution treatment to produce the γ phase, and the second is artificial ageing to produce the γ' and γ'' phases [31,32]. However, some research was performed for double artificial ageing to improve the mechanical properties of IN 718. Double ageing (DA) was used for IN 718 treatment with temperature of the specimen being 700 °C, with a holding time of 12 h for the first ageing sequence. The temperature is then reduced to 620 °C, with a hold time of 6 h as the second ageing sequence, followed by air cooling. DA increases the UTS for IN 718 to more than 1500 MPa [26,33]. IN 718 manufactured by directed energy deposition has shown the best performance of creep phenomenon after DA treatment. Its creep lifetime is 200/h, the highest value compared to other heat treatments (HT), such as HT homogenization and hot isotactic pressing and DA [34]. However, some studies explained that, without solution treatment as the first treatment in precipitation hardening of IN 718, this made the formed γ phase unstable [32].

This study fabricates IN 718 tools for a round cup deep drawing application using an LPBF printer. However, as far as the author knows, there is still no study that applies LPBF to print parts for cold deep drawing. In addition, the Taguchi method is used to determine the optimal printing parameter values for IN 718 powder that produce the best mechanical properties in terms of UTS, hardness, and surface roughness. The study then determines whether the optimal parameter values for the DA treatment for as-built IN 718 specimens produce mechanical properties that are suited to a deep drawing application. Tools are produced and set up on a stamping machine to verify their suitability in a round cup deep drawing application.

2. Materials and Methods

2.1. Research Flow

This study proposes a standard procedure to print IN 718 tools for a cold deep drawing application. The mechanical properties of the as-built, heat-treated, and surface-finished objects are determined. This study used an LPBF printer and the Taguchi method to optimize the printing parameter values for IN718 that produce the greatest UTS. These optimized printing parameters are then used to manufacture a specimen for DA treatment using different parameter values. The UTS of the printed die must be greater than 1500 MPa, and the hardness must be greater than 50 HRC for a cold deep drawing application.

Figure 1 shows the research flow for this study. Previous studies show that four factors affect the UTS: laser power, scanning speed, hatch space, and layer thickness. The printing parameter values for 9 experimental groups were determined using the Taguchi method. The specimens for each group were printed, and a tensile test was conducted. A comparison of the UTS values for the 9 groups shows the printing parameter that has the maximal UTS, and this is optimized using the result of the Taguchi method. The optimized printing parameter is used to reprint the tensile specimens. Half of the tensile specimens underwent DA treatment using different parameter values, and all specimens were subjected to a tensile test to determine the effect of DA treatment and the surface finish on the UTS and hardness. The optimal printing and DA parameter values were then used to produce tools for a deep drawing test.

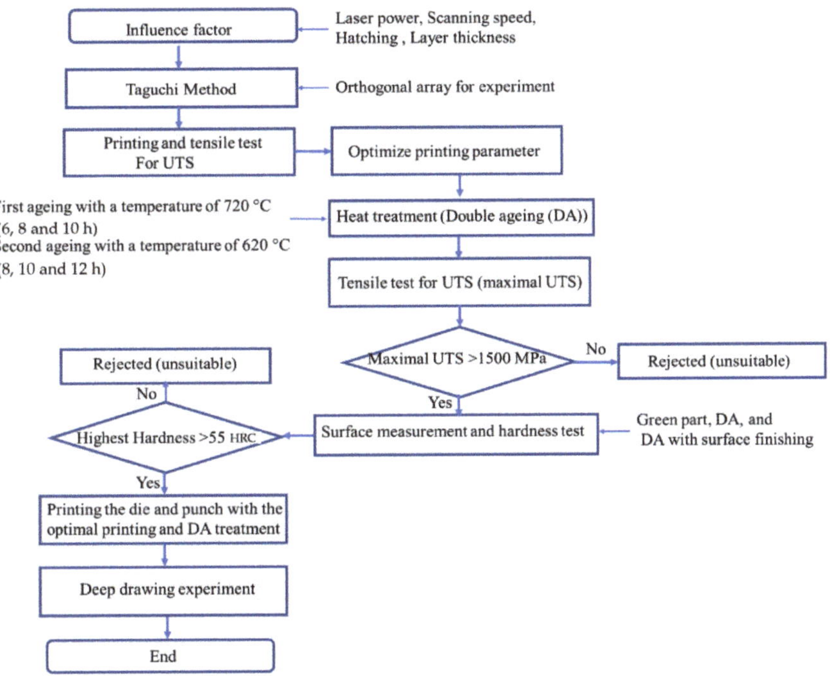

Figure 1. Research flow for this study.

2.2. Material Preparation and Laser Powder Bed Fusion Printer

IN 718 powders were purchased from Chung Yo Materials Co., Ltd., Kaohsiung, Taiwan. These were subject to sieving and drying before the printing process. The powder was sieved using a filter with a mesh size of 60 μm. The drying process used a temperature of 150 °C and a holding time of 1 h. Figure 2 shows a scanning electron microscope (SEM) image of IN 718 powder, which has a diameter of 10–50 μm.

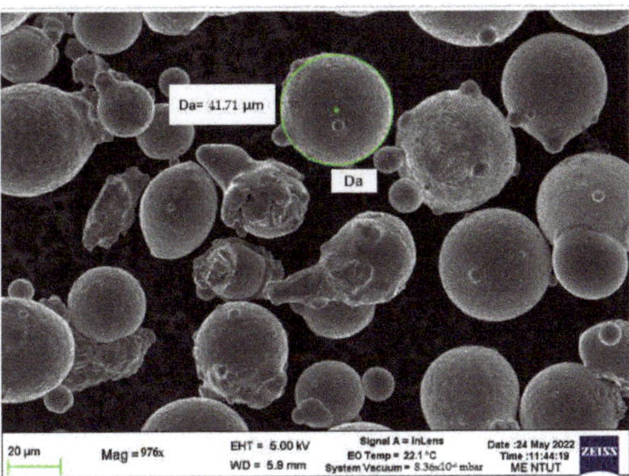

Figure 2. SEM image of IN 718 powder.

An LPBF printer (AMP-160, Tongtai Co., Kaohsiung, Taiwan), which uses selective laser melting, was used to print the specimen. The specimen model was exported using a standard tessellation language (STL) format and imported to a slicing software (Materialise Magics 23.1, Lauven, Belgia) to generate the scanning path for the printing parameter values.

2.3. Taguchi Optimization for the Printing Parameter

The Taguchi method was used to design the experiments and to determine the relationship between the printing parameters and their effect on the mechanical properties of as-built objects, including UTS and hardness. The printing parameters are laser power (P), scan speed (V), hatch spacing (H) and layer thickness (Lt), which are calculated using the volumetric laser density (VED) formula (J/mm^3) as Equation (1):

$$\text{VED} = \frac{P}{V \cdot H \cdot Lt} \tag{1}$$

This study uses the UTS as the criterion for the Taguchi method. The greater the UTS, the more optimized is the printing parameter. The signal-to-noise (S/N) ratio is a standard for quality control and is expressed as Equation (2). Increasing the S/N value decreases the standard deviation, so the parameters are more stable [19,35].

$$\frac{S}{N_i} = -10\,log\left(\frac{1}{n}\sum_{i=1}^{n}\frac{1}{y_i^2}\right) \tag{2}$$

The printed tensile specimens were subjected to tensile testing. The specimen details pertain to ASTM E8 standards. Before the printing parameters were optimized, the tensile specimen was printed using the vertical and horizontal building, as shown in Figure 3, to determine the effect of building direction difference on the tensile test result. The best build direction was used for the experiment.

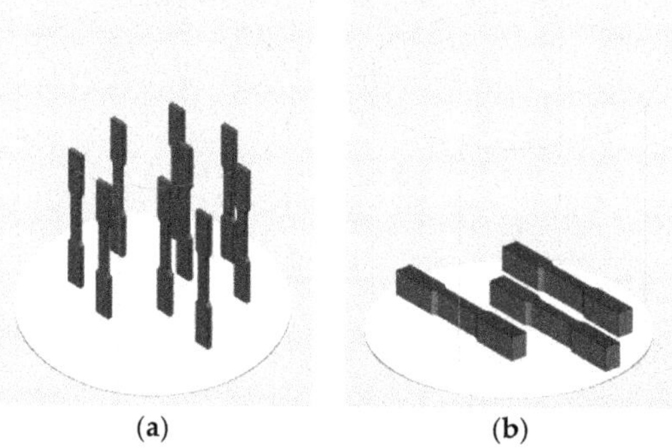

Figure 3. Tensile specimens using (**a**) vertical and (**b**) horizontal building directions.

The printing parameter values in Table 2 show that the minimum and maximum values for laser power, scanning speed, and layer thickness are 180 W and 200 W, 600 mm/s and 800 mm/s, 0.08 mm and 0.105 mm, and 30 microns and 50 microns, respectively. The Taguchi method for this study uses a level of 3, and the printing parameter table is shown in Table 2. An orthogonal L_9 table is created using Table 2 and is shown in Table 3. Each parameter in the orthogonal table occurs the same number of times, so an analysis of variance (ANOVA) is used to determine the effect of each parameter on the printing quality. A lab-developed software, based on ANOVA, was used to create the response factor plots of the average of UTS and the S/N ratio.

Table 2. Printing parameters for the Taguchi Method.

Factor Code	Parameter/Level	1	2	3
A	Laser power (P)	180	190	200
B	Scanning speed (V)	600	700	800
C	Hatch space (H)	0.08	0.09	0.105
D	Layer thickness (Lt)	30	35	40

Table 3. Orthogonal array for the experiment.

Sample Code (i)	Level P	V	H	Lt	P (W)	V (mm/s)	H (mm)	Lt (μm)	VED (J/mm³)
1	1	1	1	1	180	600	0.08	30	125
2	1	2	2	2	180	700	0.09	35	81.63
3	1	3	3	3	180	800	0.105	40	53.57
4	2	1	2	3	190	600	0.09	40	87.96
5	2	2	3	1	190	700	0.105	30	86.17
6	2	3	1	2	190	800	0.08	35	84.82
7	3	1	3	2	200	600	0.105	35	90.7
8	3	2	1	3	200	700	0.08	40	89.28
9	3	3	2	1	200	800	0.09	30	92.59

Each set of experiments was repeated 6 times ($r = 6$), and there were nine sets of experiments, generating 74 data sets. The experimental results are expressed in terms of the mean value (calculated using Equation (3)), standard deviation (calculated using Equation (4)), and S/N value (calculated using Equation (2)).

$$\overline{y_i} = \sum_{j=1}^{r} \frac{y_{ij}}{r} \tag{3}$$

$$S_i = \sqrt{\frac{\sum_{j=1}^{r} \left(y_{ij} - \overline{y_i} \right)^2}{r - 1}} \tag{4}$$

where; y_{ij} refers to the experimental data, i is the experiment of the sample code, and j is the result of the rth investigation.

2.4. Heat Treatment

IN 718 can be strengthened through precipitation hardening, and one of the types is double ageing (DA). In this study, the as-built IN 718 specimens involved DA using a furnace (HTF 1800, Carbolite Gero, Derbyshire, UK). The first ageing was carried out at a temperature of 720 °C and the second at 620 °C. The holding time in this study was varied to obtain maximum tensile strength. First ageing has holding time variations of 6, 8, and 10 h. Meanwhile, second ageing has various holding times of 8, 10, and 12 h. For more details, results can be observed in Table 4.

IN 718, printed with LPBF, according to the optimized printing parameters, was heated in the furnace until it reached 720 °C and held for a specific duration. After the holding process, the material was slowly cooled in the furnace at 55 °C per hour. The second ageing process was reheating from room temperature to 620 °C and involved holding the material at a specific time. After the holding time process, the material was cooled in the air until it reached room temperature.

Table 4. Double ageing holding times for each experiment.

Sample Code	Double Ageing Holding Time (Hour)	
	First Ageing	**Second Ageing**
T1	6	8
T2	6	10
T3	6	12
T4	8	8
T5	8	10
T6	8	12
T7	10	8
T8	10	10
T9	10	12

2.5. Hardness and Surface Roughness Measurement

Hardness was measured for the as-built, heat-treated, and surface-finished samples. Surface finishing used shot peening and polishing. The optimized printing parameter values were used to print the specimen. The hardness test used the Vickers method with ASTM E384 as the testing material standard. The hardness was measured using a Vickers hardness testing machine (HMV-G21S, Shimadzu, Kyoto, Japan). Meanwhile, surface roughness testing used visual observation with the standard of ASTM D7127. The surface roughness was measured using a three-dimensional microscope (VR300 Keyence, Osaka, Japan) with the unit of μm for the roughness average (Ra).

Figure 4a shows the die for cold deep drawing that is produced by this study. Most dies have round corners to allow the sheet material to flow into the die cavity easily, so the hardness and surface roughness are measured on the upper side, the curved side, and the edge of the die. Figure 4b shows the as-built specimens.

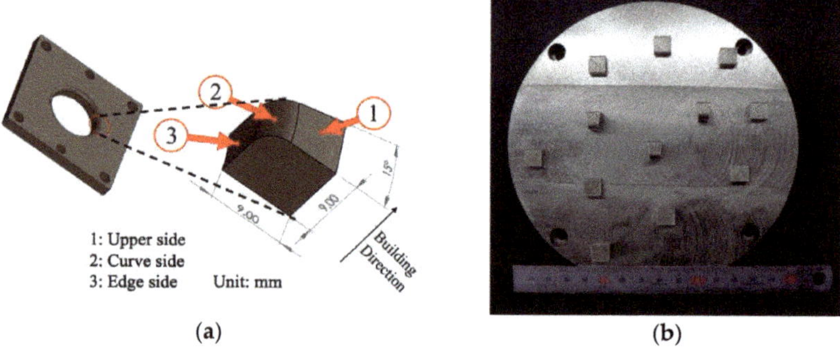

Figure 4. Hardness and surface roughness measurement points on part of die for deep drawing (**a**) and the as-built specimen (**b**).

2.6. Deep Drawing Application

A die and a punch were produced for a cup deep drawing experiment using the optimized printing and DA parameter values in order to determine the suitability of the material to a deep cold drawing application. A simple round cup shape was used to eliminate complication in the validation of the process, as shown in Figure 5a. The post-processed die was assembled using the blank holder, positioning pins, and a bottom plate. The assembled die set was fixed on a mounting table. The post-processed punch was fixed to the punch holder and driven using a press ram. An aluminum alloy (Al) 6061 sheet with a thickness of 1 mm was placed between the post-processed die and the blank holder.

The press ram drove the punch downward to force the sheet to flow into the die to form a round cup.

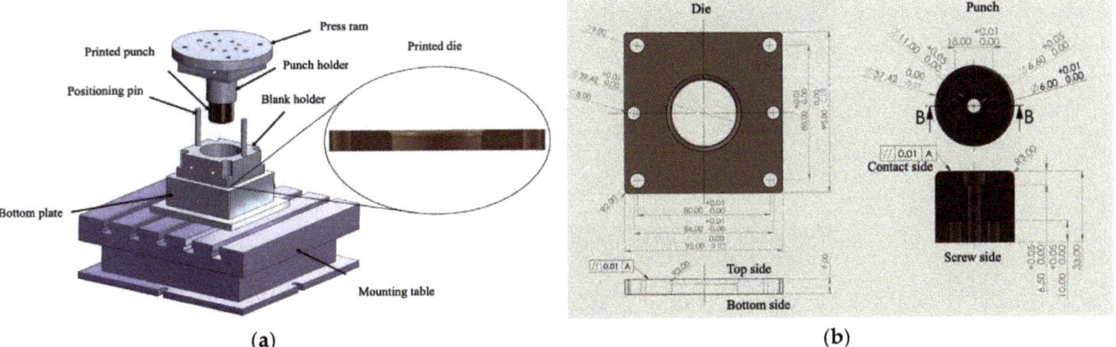

Figure 5. Schematic diagram of a deep drawing die set (**a**) and the dimensions of the die and punch (**b**).

Figure 5b shows the dimensions of the die and punch. The die is $95 \times 95 \times 9$ mm^3. The center of the die has a hole with a diameter of 39.42 mm. The diameter and height of punch are 37.42 mm and 33 mm, respectively. The edges of the hole on the top side and the punch on the contact side have a radius of 3 mm. The thickness axis is the building direction for the LPBF printer.

3. Results and Discussion

3.1. Processing Parameters Optimization by the Taguchi Method

Specimens were printed in a horizontal and vertical orientation and were subjected to a tensile test, as shown in Figure 6. The results of the tensile test and the S/N ratio are listed in Table 5. The average ultimate strength (UTS) values for the specimens that are printed using a vertical building direction are less than those that are printed using a horizontal building direction. The S/N ratio for the horizontally printed specimens is greater than 60, and the smallest average UTS is 1046.31 MPa. The printing optimization and heat treatment tests use horizontally printed specimens. This result is supported by previous research [36] as a material printed with PBLF, which has a higher tensile strength for the building direction of horizontal, rather than vertical.

Table 5. L_9 orthogonal result for the tensile test and the S/N ratio.

Sample Code (*i*)	Vertical Direction		Horizontal Direction	
	Average UTS (MPa)	*S/N*	**Average UTS (MPa)**	*S/N*
1	942.72	59.87	1078.09	60.65
2	1028.75	60.24	1046.31	60.39
3	1037.4	60.32	1071.47	60.6
4	1033.01	60.28	1091.33	60.76
5	989.77	59.91	1073.11	60.61
6	1003.2	60.03	1048.41	60.41
7	1055.68	60.47	1056.63	60.45
8	1012.24	60.1	1056.8	60.48
9	977.78	59.8	1050.18	60.42

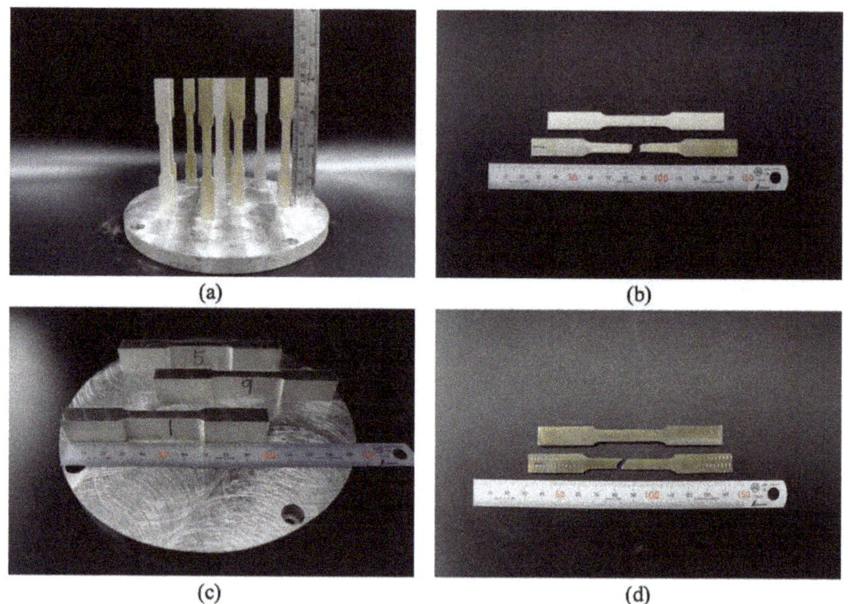

Figure 6. Photographs of the as-built specimens and the specimen after the tensile test: (**a**,**b**) are printed in the vertical building direction, and (**c**,**d**) are in the horizontal building direction.

To optimize the printing parameters, the response graphs for average UTS and S/N ratio are plotted in Figure 7. Laser power, scanning speed, hatch space, and layer thickness are, respectively, denoted as A, B, C, and D. Figure 7a shows that the order in which the response factors affect the UTS results for printed IN718 is: layer thickness, hatch space, scanning speed, and laser power. This result contrasts with the results of another study [23] because each study uses a different printing mechanism. The previous study used a six-axis robot with a fiber laser and a powder feeder system to deposit the powder on the laser focus zone. The printing mechanism for this study paves the powder on the platform and deposits the laser energy selectively along the scanning path to induce a phase change in the powder from the solid state to the liquid state.

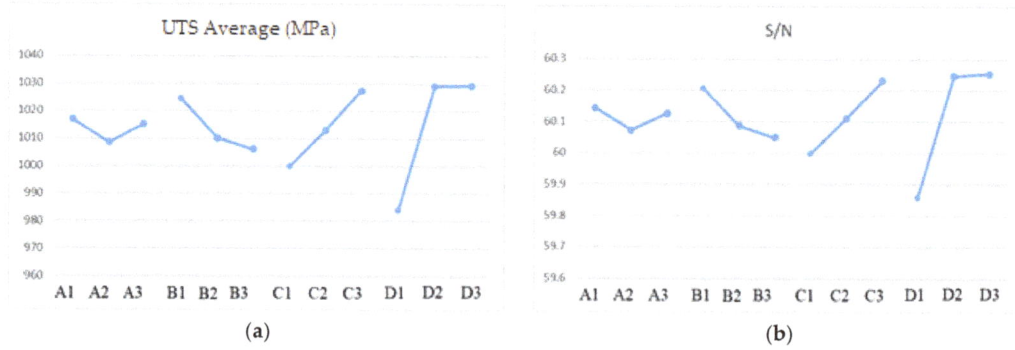

Figure 7. Factor response plot for (**a**) the average UTS and (**b**) the S/N ratio.

The S/N ratio plot in Figure 7b shows that an increase in the hatch spacing (C) and layer thickness (D) increases the UTS for the as-built object. The powder has an average particle size of 10–50 µm, so the layer must be thicker than 40 µm. The hatch space for Sample 4 is increased to 0.105 mm from 0.09 mm, and this is denoted as Sample 10. For

6 test pieces that were printed using a horizontal building direction, the highest UTS value is 1122.88 MPa. The optimal values for the printing parameters are listed in Table 6, and the volumetric laser density formula is 75.4 J/mm^3. The greatest value for UTS is less than 1500 MPa [26], so heat treatment is required. The as-built specimens for this experiment use the printing parameters for Sample 10.

Table 6. The optimal values for printing parameters for IN 718, as defined using the Taguchi method.

Sample Code (i)	Printing Parameter				Experimental Result	S/N	VED_H (J/mm^3)
	P	V	H	Lt	UTS (MPa)		
10	190	600	0.105	40	1122.88	60.58	75.4

3.2. Heat Treatment

The minimum respective values for UTS and hardness for cold deep drawing tools are 1500 MPa and 50 HRC [28,29]. The maximum UTS value for Sample 10 is 1122.88 MPa, which is less than 1500 MPa, so DA treatment was used to increase the UTS for the as-built object using the design in Table 4. The UTS after DA treatment is listed in Table 4 and shown in the Figure 8. The experimental results show that the UTS for the T1, T2, T3, and T4 groups is significantly greater than 1500 MPa, and T4 features the highest UTS. Therefore, the optimal DA treatment involves heating the printed specimens to 720 °C at 10 °C per minute and maintaining this temperature for 8 h and then cooling in the furnace at a cooling rate of 55 °C per hour to a temperature of 620 °C and maintaining this temperature for 8 h.

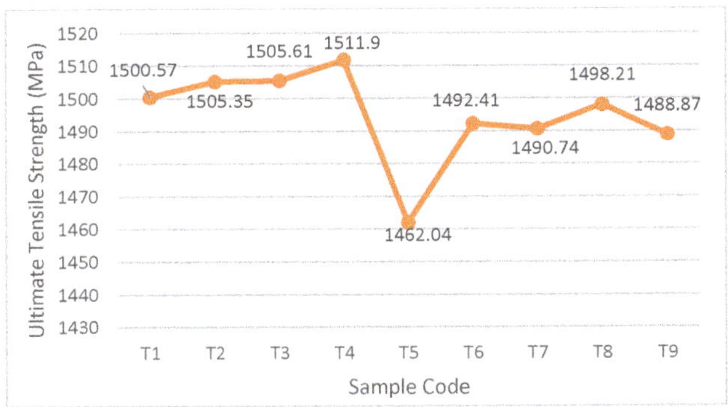

Figure 8. The maximum UTS for the tensile test for the specimen that is subject to DA treatment.

Based on Figure 8, T5 shows the lowest UTS of IN 718 value while the duration of the first and second artificial ageing holding time is neither the shortest nor the longest duration. This process is due to the unstable γ phase in multiple ageing. Some elements that cannot be released during the γ formation process, such as niobium, titanium, and molybdenum, result in other phases not being formed stably or even not being formed. This has an impact on reducing the strength of IN 718 [31].

3.3. Surface Finishing and Hardness Measurement

Figure 9 shows the results for surface roughness. The surface roughness of the as-built specimen is greatest on the upper side, which has a value of 11.72 μm for the Ra, but the upper side becomes slightly smoother after DA treatment. The surface roughness of the curved side is greatest because it must have a staircase effect, but experimental results show that it is less than the value for the upper side, possibly because the layer is thinner, so the staircase effect is eliminated [37]. After DA treatment and surface finishing, the surface roughness is reduced to 2 μm for the Ra.

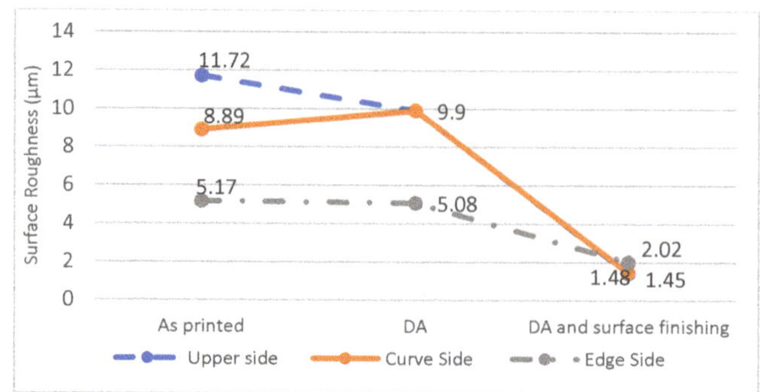

Figure 9. Surface roughness measurement for the as-built specimens, specimens that are subject to DA, and post-processed specimens (subject to DA and surface finishing).

Figure 10 shows the Rockwell hardness results. Figure 10a shows that the distance between each indentation point on the test object is 4 mm in order to avoid measurement errors when the material changes phase. Figure 10b shows that the respective hardness of the as-built specimens, specimens that are subject to DA, and post-processed specimens, which include HRC 32, HRC 46, and HRC 55.

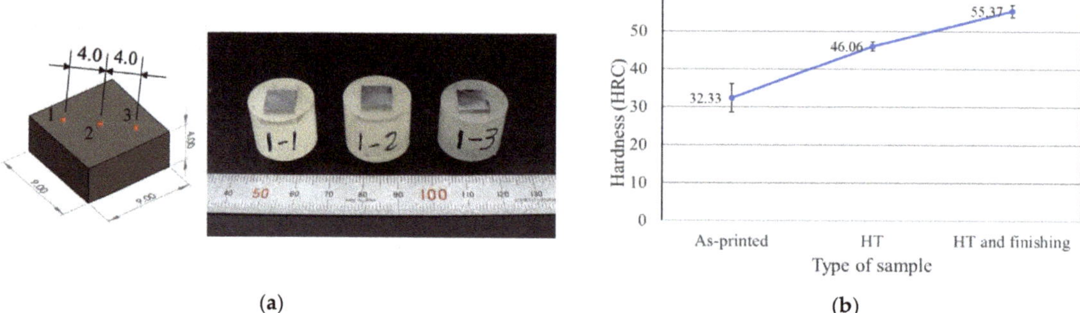

(a)   (b)

Figure 10. Rockwell hardness measurement: (**a**) the distance between each indentation point and (**b**) results for the as-built specimens, specimens that are subject to DA treatment, and post-processed specimens.

Amato et al. annealed an as-built specimen at 982 °C for 0.5 h under vacuum and then used a hot isostatic pressing process (HIP) at 1163 °C and 0.1 GPa pressure for 4 h in argon. The maximum hardness of the as-built and annealed with HIP specimens is 33 HRC and 38 HRC, respectively [38]. Compared to it, the hardness of the treated specimen for this study is greater than 38 HRC and increases to 55 HRC after surface finishing. This reveals that the optimal parameter of this study can approach to practical application.

3.4. Deep Drawing Verification

Figures 11a and 11b, respectively, show the as-built die and punch on the platform. Both were subject to DA treatment and were surface polished using wire-electrical discharge machining (WEDM) after they were removed from the platform. The dimensions of both comply with the specifications, and they were assembled on the punch holder and die set, as shown in Figure 11c,d. Figure 12 (left) shows a sheet on the die that is clamped using the blank holder. After deep drawing, a round cup was formed, as shown in Figure 12 (right). The shape of the round cup is simple, but it verifies that the optimized printing and DA treatment parameter values produce a punch and die that are suited for use in a cold deep drawing application.

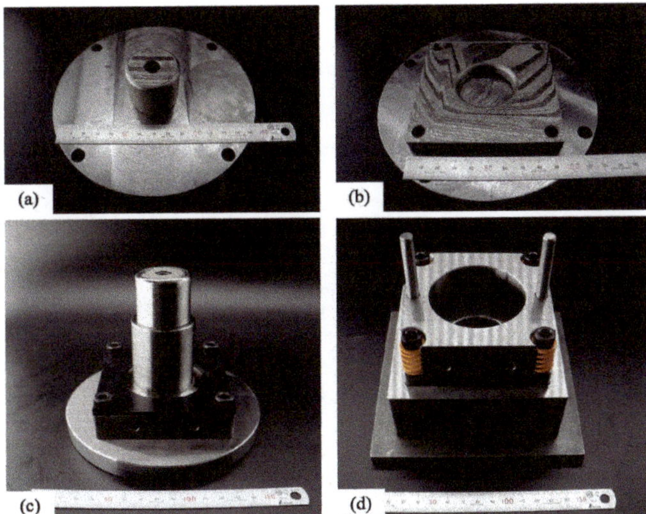

Figure 11. Photographs of the as-built punch (**a**) and die (**b**) on the platform and the assembled punch set (**c**) and die set (**d**).

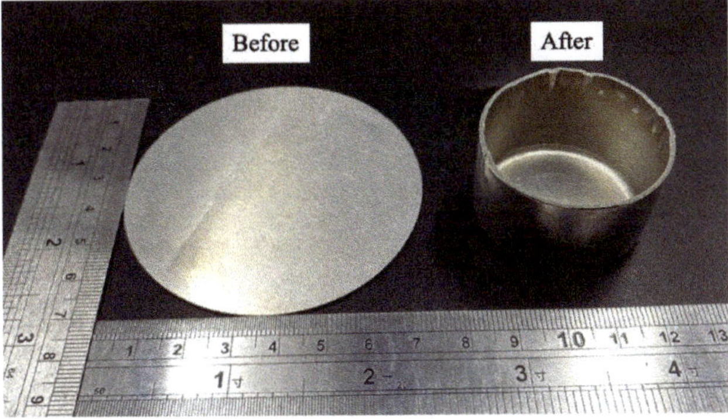

Figure 12. Photograph of the Al 6061 sheet (**left**) and the drawn round cup (**right**).

This study optimizes the printing parameter values for IN 718 and the post-processing parameter values for the as-built parts to create mechanical properties that render this material suited to the production of a cold deep drawing die, but service life for the round cup deep drawing process is not considered, so future studies will optimize the topological structure of as-built die to eliminate material waste and increase the service life of the post-processed die.

4. Conclusions

This study optimizes the values for the printing and heat treatment parameters for IN 718 to produce the tools that are suited for use in a cold deep drawing application. The Taguchi method is used to determine the effect of the printing parameters on the mechanical properties of the as-built object. The following conclusions are drawn:

1. The results of the Taguchi method show that the order in which the relevant factors affect LPBF printing is: layer thickness, hatch space, scanning speed, and laser power. However, changing the value for hatch space has the most significant effect because the diameter of the powder particles defines the least thickness for each layer.

2. The optimized printing parameter values include a laser power of 190 W, a scanning speed of 600 mm/s, a hatch space of 0.105 mm, and a layer thickness of 40 μm to produce a maximum UTS of 1122.88 MPa. The hardness of the as-built specimen is 32.33 HRC.

3. The optimal parameters for heat treatment are a temperature of 720 °C with a holding time of 8 h for the first ageing sequence, a decrease in temperature to 620 °C with a holding time of 8 h for the second sequence, and cooling in the furnace at a cooling rate of 55 °C per hour. After heat treatment, the UTS increases to 1511.9 MPa, and the hardness increases to 46.06 HRC. After surface finishing, the hardness increases to 55.37 HRC.

4. The optimized values for the printing and heat treatment parameters give a tensile strength of more than 1500 MPa and a hardness of more than 50 HRC, which meet the requirements for a tool for a cold deep drawing application.

5. The results of the deep drawing experiment verify that the optimized values for the printing and post-processing parameters produce a die and punch that form an Al 6061 round cup.

6. Although the process of optimizing printing parameters and double aging produces materials that are suitable for standard die parts for cold deep drawing, judging from the results of the Taguchi calculations, the results are in the unsatisfactory category and can be further improved with advanced Taguchi analysis.

7. In terms of material, there is still much that can be explored for the application of IN 718, which is printed using LPBF as a cold deep drawing dies part, such as post-printing material characterization, mechanical behavior, fatigue and failure behavior, and many others. Optimizing printing parameters using other parameters is still very possible to do in the future. This is because the printing parameters are not only related to the four parameters that we mention in this study.

Author Contributions: Conceptualization, C.-P.J. and A.A.M.; methodology, C.-P.J., A.A.M. and S.-H.W.; validation, A.A.M., S.-H.W. and Y.-R.W.; investigation, C.-P.J.; resources, C.-P.J.; data curation, A.A.M., M.R. and Y.-R.W.; writing—original draft preparation, C.-P.J., A.A.M. and T.P.; writing—review and editing, M.R. and T.P.; supervision, C.-P.J.; project administration, C.-P.J.; funding acquisition, C.-P.J. All authors have read and agreed to the published version of the manuscript.

Funding: The authors are grateful for financial support by the Ministry of Science and Technology (MOST), Taiwan, under Grant Nos. 111-2622-E-027-028 and 111-2221-E-027-087-MY2.

Institutional Review Board Statement: Not applicable.

Informed Consent Statement: Not applicable.

Data Availability Statement: The data presented in this study are available upon request from the corresponding author.

Acknowledgments: This work also supported by the High-Value Biomaterials Research and Commercialization Center of the Featured Area Research Center Program, within the framework of the Higher Education Sprout Project of the Taiwanese Ministry of Education (MOE).

Conflicts of Interest: The authors declare no conflict of interest.

References

1. Lu, C.; Shi, J. Simultaneous consideration of relative density, energy consumption, and build time for selective laser melting of IN 718: A multi-objective optimization study on process parameter selection. *J. Clean. Prod.* **2022**, *369*, 133284. [CrossRef]

2. Sing, S.L.; Yeong, W.Y. Laser powder bed fusion for metal additive manufacturing: Perspectives on recent developments. *Virtual Phys. Prototyp.* **2020**, *15*, 359–370. [CrossRef]

3. Oliveira, J.P.; LaLonde, A.; Ma, J. Processing parameters in laser powder bed fusion metal additive manufacturing. *Mater. Design* **2020**, *193*, 108762. [CrossRef]

4. Petroušek, P.; Kvačkaj, T.; Bidulská, J.; Bidulský, R.; Grande, M.A.; Manfredi, D.; Pokorný, I. Investigation of the Properties of 316L Stainless Steel after AM and Heat Treatment. *Materials* **2023**, *16*, 3935. [CrossRef]

5. Nandhakumar, R.; Venkatesan, K. A process parameters review on Selective laser melting-based additive manufacturing of Single and Multi-Material: Microstructure, Properties, and machinability aspects. *Mater. Today Commun.* **2023**, *35*, 105538. [CrossRef]

6. Wang, W.; Wang, S.; Zhang, X.; Chen, F.; Xu, Y.; Tian, Y. Process parameter optimization for selective laser melting of IN 718 superalloy and the effects of subsequent heat treatment on the microstructural evolution and mechanical properties. *J. Manuf. Process.* **2021**, *64*, 530–543. [CrossRef]

7. Baldi, N.; Giorgetti, A.; Palladino, M.; Giovannetti, I.; Arcidiacono, G.; Citti, P. Study on the Effect of Inter-Layer Cooling Time on Porosity and Melt Pool in Inconel 718 Components Processed by Laser Powder Bed Fusion. *Materials* **2023**, *16*, 3920. [CrossRef]

8. Zhang, D.; Niu, W.; Cao, X.; Liu, Z. Effect of standard heat treatment on the microstructure and mechanical properties of selective laser melting manufactured IN 718 superalloy. *Mater. Sci. Eng. C.* **2015**, *644*, 32–40. [CrossRef]

9. Xi, N.; Fang, X.; Duan, Y.; Zhang, Q.; Huang, K. Wire arc additive manufacturing of IN 718: Constitutive modelling and its microstructure basis. *J. Manuf. Process.* **2022**, *75*, 1134–1143. [CrossRef]

10. Wang, H.; Wang, L.; Cui, R.; Wang, B.; Luo, L.; Su, Y. Differences in microstructure and nano-hardness of selective laser melted IN 718 single tracks under various melting modes of molten pool. *J. Mater. Res. Technol.* **2020**, *9*, 10401–10410. [CrossRef]

11. Kaletsch, A.; Qin, S.; Broeckmann, C. Influence of Different Build Orientations and Heat Treatments on the Creep Properties of Inconel 718 Produced by PBF-LB. *Materials* **2023**, *16*, 4087. [CrossRef] [PubMed]

12. Schneider, J.; Lund, B.; Fullen, M. Effect of heat treatment variations on the mechanical properties of IN 718 selective laser melted specimens. *Addit. Manuf.* **2018**, *21*, 248–254. [CrossRef]

13. Lesyk, D.; Martinez, S.; Mordyuk, B.; Dzhemelinskyi, V.; Lamikiz, A.; Prokopenko, G. Post-processing of the IN 718 alloy parts fabricated by selective laser melting: Effects of mechanical surface treatments on surface topography, porosity, hardness and residual stress. *Surf. Coat. Technol.* **2020**, *381*, 125136. [CrossRef]

14. Baicheng, Z.; Xiaohua, L.; Jiaming, B.; Junfeng, G.; Pan, W.; Chen-nan, S.; Muiling, N.; Guojun, Q.; Jun, W. Study of selective laser melting (SLM) IN 718 part surface improvement by electrochemical polishing. *Mater. Design* **2017**, *116*, 531–537. [CrossRef]

15. Xu, Z.; Cao, L.; Zhu, Q.; Guo, C.; Li, X.; Hu, X.; Yu, Z. Creep property of IN 718 superalloy produced by selective laser melting compared to forging. *Mater. Sci. Eng. C.* **2020**, *794*, 139947. [CrossRef]

16. Godec, M.; Malej, S.; Feizpour, D.; Donik, Č.; Balažic, M.; Klobčar, D.; Pambaguian, L.; Conradi, M.; Kocijan, A. Hybrid additive manufacturing of IN 718 for future space applications. *Mater. Charact.* **2021**, *172*, 110842. [CrossRef]

17. McLouth, T.D.; Bean, G.E.; Witkin, D.B.; Sitzman, S.D.; Adams, P.M.; Patel, D.N.; Park, W.; Yang, J.-M.; Zaldivar, R.J. The effect of laser focus shift on microstructural variation of IN 718 produced by selective laser melting. *Mater. Design* **2018**, *149*, 205–213. [CrossRef]

18. Balbaa, M.; Mekhiel, S.; Elbestawi, M.; McIsaac, J. On selective laser melting of IN 718: Densification, surface roughness, and residual stresses. *Mater. Design* **2020**, *193*, 108818. [CrossRef]

19. Jiang, H.-Z.; Li, Z.-Y.; Feng, T.; Wu, P.-Y.; Chen, Q.-S.; Feng, Y.-L.; Li, S.-W.; Gao, H.; Xu, H.-J. Factor analysis of selective laser melting process parameters with normalised quantities and Taguchi method. *Opt. Laser Technol.* **2019**, *119*, 105592. [CrossRef]

20. Xiansheng, N.; Zhenggan, Z.; Xiongwei, W.; Luming, L. The use of Taguchi method to optimize the laser welding of sealing neuro-stimulator. *Opt. Lasers Eng.* **2011**, *49*, 297–304. [CrossRef]

21. Sheshadri, R.; Nagaraj, M.; Lakshmikanthan, A.; Chandrashekarappa, M.P.G.; Pimenov, D.Y.; Giasin, K.; Prasad, R.V.S.; Wojciechowski, S. Experimental investigation of selective laser melting parameters for higher surface quality and micro-hardness properties: Taguchi and super ranking concept approaches. *J. Mater. Res. Technol.* **2021**, *14*, 2586–2600. [CrossRef]

22. Jiang, C.-P.; Cheng, Y.-C.; Lin, H.-W.; Chang, Y.-L.; Pasang, T.; Lee, S.-Y. Optimization of FDM 3D printing parameters for high strength PEEK using the Taguchi method and experimental validation. *Rapid Prototyp. J.* **2022**, *28*, 1260–1271. [CrossRef]

23. Yang, B.; Lai, Y.; Yue, X.; Wang, D.; Zhao, Y. Parametric optimization of laser additive manufacturing of Inc 625 using Taguchi method and grey relational analysis. *Scanning* **2020**, *2020*, 9176509. [CrossRef] [PubMed]

24. Sheheryar, M.; Khan, M.A.; Jaffery, S.H.I.; Alruqi, M.; Khan, R.; Bashir, M.N.; Petru, J. Multi-Objective Optimization of Process Parameters during Micro-Milling of Nickel-Based Alloy Inconel 718 Using Taguchi-Grey Relation Integrated Approach. *Materials* **2022**, *15*, 8296. [CrossRef] [PubMed]

25. Maicas-Esteve, H.; Taji, I.; Wilms, M.; Gonzalez-Garcia, Y.; Johnsen, R. Corrosion and Microstructural Investigation on Additively Manufactured 316L Stainless Steel: Experimental and Statistical Approach. *Materials* **2022**, *15*, 1605. [CrossRef] [PubMed]

26. Huang, W.; Yang, J.; Yang, H.; Jing, G.; Wang, Z.; Zeng, X. Heat treatment of IN 718 produced by selective laser melting: Microstructure and mechanical properties. *Mater. Sci. Eng. C* **2019**, *750*, 98–107. [CrossRef]

27. Stolt, R.; André, S.; Elgh, F. Introducing inserts for die casting manufactured by selective laser sintering. *Procedia Manuf.* **2018**, *17*, 309–316. [CrossRef]

28. Li, Z.X.; Shi, Y.L.; Xu, L.P.; Jin, J.X. Effect of Natural Aging on Cold Forming Performance of 2219 Aluminum Alloy. *Materials* **2023**, *16*, 3536. [CrossRef]

29. Do, T.T.; Minh, P.S.; Le, N. Effect of tool geometry parameters on the formability of a camera cover in the deep drawing process. *Materials* **2021**, *14*, 3993. [CrossRef]

30. Yan, J.; Zheng, D.; Li, H.; Jia, X.; Sun, J.; Li, Y.; Qian, M.; Yan, M. Selective laser melting of H13: Microstructure and residual stress. *J. Mater. Sci.* **2017**, *52*, 12476–12485. [CrossRef]

31. Chen, W.; Chaturvedi, M.; Richards, N. Effect of boron segregation at grain boundaries on heat-affected zone cracking in wrought IN 718. *Metall. Mater. Trans. A* **2001**, *32*, 931–939. [CrossRef]

32. Chamanfar, A.; Sarrat, L.; Jahazi, M.; Asadi, M.; Weck, A.; Koul, A. Microstructural characteristics of forged and heat treated Inconel-718 disks. *Mater. Des.* **2013**, *52*, 791–800. [CrossRef]

33. Teng, Q.; Li, S.; Wei, Q.; Shi, Y. Investigation on the influence of heat treatment on IN 718 fabricated by selective laser melting: Microstructure and high temperature tensile property. *J. Manuf. Process.* **2021**, *61*, 35–45. [CrossRef]

34. Li, Y.; Podaný, P.; Koukolíková, M.; Džugan, J.; Krajňák, T.; Veselý, J.; Raghavan, S. Effect of Heat Treatment on Creep Deformation and Fracture Properties for a Coarse-Grained Inconel 718 Manufactured by Directed Energy Deposition. *Materials* **2023**, *16*, 1337. [CrossRef] [PubMed]

35. Javidi, M.; Hosseini, S.; Khodabakhshi, F.; Mohammadi, M.; Orovcik, L.; Trembošová, V.N.; Nagy, Š.; Nosko, M. Laser powder bed fusion of 316L stainless steel/Al2O3 nanocomposites: Taguchi analysis and material characterization. *Opt. Laser Technol.* **2023**, *158*, 108883. [CrossRef]

36. Pilgar, C.M.; Fernandez, A.M.; Lucarini, S.; Segurado, J. Effect of printing direction and thickness on the mechanical behavior of SLM fabricated Hastelloy-X. *Int. J. Plast.* **2022**, *153*, 3250. [CrossRef]

37. Chantzis, D.; Liu, X.; Politis, D.J.; El Fakir, O.; Chua, T.Y.; Shi, Z.; Wang, L. Review on additive manufacturing of tooling for hot stamping. *Int. J. Adv. Manuf. Technol.* **2020**, *109*, 87–107. [CrossRef]

38. Amato, K.; Gaytan, S.; Murr, L.E.; Martinez, E.; Shindo, P.; Hernandez, J.; Collins, S.; Medina, F. Microstructures and mechanical behavior of IN 718 fabricated by selective laser melting. *Acta Mater.* **2012**, *60*, 2229–2239. [CrossRef]

Article

Enhanced Hardness-Toughness Balance Induced by Adaptive Adjustment of the Matrix Microstructure in In Situ Composites

Mingjuan Zhao [1,2], Xiang Jiang [1,2], Yumeng Guan [1,2], Haichao Yang [1,2], Longzhi Zhao [1,2,3,*], Dejia Liu [1,2,3], Haitao Jiao [1,2,3], Meng Yu [4], Yanchuan Tang [1,2,3,*] and Laichang Zhang [5]

1   Key Laboratory of Advanced Materials for Vehicles and Laser Additive Manufacturing of Nanchang City, East China Jiaotong University, Nanchang 330013, China
2   Key Laboratory of Vehicle and Equipment of Education Ministry, East China Jiaotong University, Nanchang 330013, China
3   State Key Laboratory of Performance Monitoring and Protecting of Rail Transit Infrastructure, East China Jiaotong University, Nanchang 330013, China
4   Nanchang Railway Tongda Industry and Trade Co., Ltd., Nanchang 330026, China
5   School of Engineering, Edith Cowan University, 270 Joondalup Drive, Joondalup, Perth, WA 6027, Australia
*   Correspondence: zhaolongzhi@163.com (L.Z.); tangyanchuan89@163.com (Y.T.); Tel.: +86-79187046559 (L.Z.); +86-79187046718 (Y.T.)

Abstract: With the development of high-speed and heavy-haul railway transportation, the surface failure of rail turnouts has become increasingly severe due to insufficient high hardness-toughness combination. In this work, in situ bainite steel matrix composites with WC primary reinforcement were fabricated via direct laser deposition (DLD). With the increased primary reinforcement content, the adaptive adjustments of the matrix microstructure and in situ reinforcement were obtained at the same time. Furthermore, the dependence of the adaptive adjustment of the composite microstructure on the composites' balance of hardness and impact toughness was evaluated. During DLD, the laser induces an interaction among the primary composite powders, which leads to obvious changes in the phase composition and morphology of the composites. With the increased WC primary reinforcement content, the dominant sheaves of the lath-like bainite and the few island-like retained austenite are changed into needle-like lower bainite and plenty of block-like retained austenite in the matrix, and the final reinforcement of Fe_3W_3C and WC is obtained. In addition, with the increased primary reinforcement content, the microhardness of the bainite steel matrix composites increases remarkably, but the impact toughness decreases. However, compared with conventional metal matrix composites, the in situ bainite steel matrix composites manufactured via DLD possess a much better hardness-toughness balance, which can be attributed to the adaptive adjustment of the matrix microstructure. This work provides a new insight into obtaining new materials with a good combination of hardness and toughness.

Keywords: in situ bainite steel matrix composite; direct laser deposition; adaptive adjustment of matrix microstructure; good hardness-toughness balance

Citation: Zhao, M.; Jiang, X.; Guan, Y.; Yang, H.; Zhao, L.; Liu, D.; Jiao, H.; Yu, M.; Tang, Y.; Zhang, L. Enhanced Hardness-Toughness Balance Induced by Adaptive Adjustment of the Matrix Microstructure in In Situ Composites. *Materials* **2023**, *16*, 4437. https://doi.org/10.3390/ma16124437

Academic Editors: Aivaras Kareiva and Damon Kent

Received: 21 April 2023
Revised: 29 May 2023
Accepted: 6 June 2023
Published: 16 June 2023

1. Introduction

With the development of high-speed and heavy-haul railway transportation, rolling-contact fatigue crack and peeling on the surface of rail turnouts have become increasingly severe [1–3]. The surface failure of rail turnouts severely reduces their service life, which leads to increased operation costs and potential safety hazards [4]. It is well known that the failure of rail turnouts is closely related to the insufficient hardness and toughness of the components [5]. Hence, it is of great importance to develop a rail turnout material with high hardness and high toughness.

Compared with the conventional surface treatment methods (such as thermal spraying, plasma spraying), laser surface-treatment technologies demonstrate obvious advantages, such as a small heat effect zone, good interface bonding, high reliability and high precision [6,7]. Specifically, during direct laser deposition (DLD), a high-energy laser beam is used to melt the composite powder, and then the melts are deposited on the substrate surface to form the bulk deposition. During the DLD process, the composition and microstructure of the deposition can be regulated [8], which makes the DLD technology suitable for strengthening the surface of the materials. For example, martensite Fe-based alloys with good wear resistance can be deposited on the surface of rail components via DLD [9–15]. However, martensite Fe-based alloys often demonstrate insufficient toughness [9], and cannot meet the demand of the rail component suffering impact load. Fe-based alloys with a bainite microstructure have a good balance of wear resistance and toughness; therefore, isothermal heat treatment is used for Fe-based alloys in order to change their martensite microstructure into bainite, but the heat treatment needs a long time, which cannot meet the low production cycle of the rail component. Furthermore, the wear resistance of bainite steel is not high enough to suit rail turnouts [13,15].

It has been reported that metal matrix composites (MMCs) exhibit excellent mechanical properties, such as a high modulus, high strength and good wear resistance, compared with their primary alloy matrix [16–18]. Thus, MMCs with a high volume fraction reinforcement are required in order to enhance the wear resistance of rail turnouts [19,20]. However, the toughness of MMCs with high volume fraction reinforcement is low [21–23]. Consequently, it is difficult to obtain a balance between high toughness and good wear resistance in conventional MMCs [24–27].

In conventional MMCs manufactured via powder metallurgy, their casting, matrix microstructure and composition are almost not changed compared with their primary alloy matrix. Therefore, the enhancement of the composites' performance is mainly attributed to the strengthening of the reinforcement, but has nothing to do with the matrix. However, during laser deposition, the complex interaction between the deposited composite powder irradiated by the high-energy-density laser induces the in situ action and the solution in the deposited powder, thus making the matrix constituent and microstructure change with the increased primary reinforcement volume fraction.

Al Mangour [28–31] suggested that that the particle reinforcement with relatively small size would melt during manufacturing TiC reinforced 316L stainless steel matrix composites by selective laser melting(SLM). Furthermore, the content of the ferrite phase (α-Fe) in the matrix increases with the increased TiC reinforcement content. The melting of the particle reinforcement, irradiated by a high-energy laser beam, has also been widely reported in the literature [32–34]. This phenomenon leads to a large deviation in the microstructure and properties of the composites without the interaction between the primary matrix and reinforcement.

Accordingly, the interaction between the primary matrix and reinforcement can also be utilized to regulate the microstructure of the composite matrix. In our previous work, a bainite steel matrix composite was fabricated, and the decomposition or dissolution of the reinforcing particles in the matrices were also found due to the extremely high temperature during the DLD process, which changes the chemical constituent and microstructure of the matrix [35–37]. Furthermore, owing to the change in the matrix constituent, the content of retained austenite in the matrix was found to vary in a large range spontaneously. Therefore, the adaptive change in the matrix comes from the content of primary particle reinforcement, which provides a much more convenient way to regulate the toughness of the composites.

Due to the combination of a high hardness and low thermal expansion coefficient, WC particle reinforcement is often used in Fe-based composites [38]. In addition, the dissolution of the W and C elements derived from WC particles exert a significant influence on the transformation of undercooled austenite, which can be used to regulate the microstructure of the Fe-based alloy matrix [39]. In this work, WC particles were used as the primary reinforcement, and the effect of the volume fraction of primary reinforcement on the matrix microstructure and mechanical properties of the in situ bainite steel matrix composite was investigated. Finally, the dependence of the adaptive adjustment of the matrix on the composites' balance of hardness and impact toughness was evaluated. This work provides new insights into obtaining new materials with a good combination of hardness and toughness.

2. Materials and Methods

The bainite steel matrix composite was fabricated via DLD on a U75V steel substrate (a kind of railway material). The surface of the substrate was ground and then sand blasted in order to remove the surface oxide layer. The gas-atomized Fe-based alloyed powder with a particle size of 50~70 μm in diameter was applied to construct the matrix of the composite. The chemical compositions (in wt%) of the substrate and Fe-based alloyed powder are shown in Table 1. In order to avoid the sputtering of the tungsten carbide ceramic particle from the composite powder by the laser during laser deposition, tungsten carbide coated with a Co layer was used as a primary reinforcement. The composite powder containing the Fe-based powder and WC powder were thoroughly mixed using a planetary ball mill in an argon atmosphere at a speed of 200 rpm for 2 h. Finally, the composite powders were dried in a vacuum furnace at 80 °C for 2 h.

Table 1. Chemical composition (in wt%) of the Fe-based powder and U75V steel substrate.

	C	Si	Mn	Cr	Ni	Mo	Al	V	Fe
Substrate	0.78	0.66	0.96	/	/	/	/	0.05	Bal.
Powder	0.45	0.90	1.20	0.90	1.90	0.30	1.20	/	Bal.

The bainite steel matrix composites were manufactured using a laser processing system (as shown in Figure 1) comprising a semiconductor laser device with a maximum output power of 2.5 kW (LDM-2500-60, Laserline, Mülheim-Kärlich, Germany), a three-axis numerical control machine controlling the laser scanning path, a powder coaxial nozzle feeding system with a shielding gas device and a stable temperature platform. The process involved the following 3 steps. Firstly, the substrates were heated to 300 ± 5 °C in the resistance furnace using argon protection and then placed on a platform with the pre-set heated temperature of 300 °C to avoid martensite transformation during laser deposition. Afterwards, the composite powder was deposited on the surface of the substrates using DLD technology. The processing parameters were as follows: laser power of 800 W, laser spot diameter of 1.5 ± 0.1 mm, overlap ratio of 40% and scanning velocity of 360 mm/min. As shown in Figure 1b, the samples had a good surface quality and no macroscopic cracks were observed. Finally, the composite samples were put into a 300 ± 3 °C salt bath for isothermal treatment for 200 min, and the final composites, air cooled to room temperature (RT), were obtained. Specimens were sectioned using electric discharge wire cutting to obtain the composite samples and characterize their microstructure and mechanical properties.

An optical microscope (OM, Carl Zeiss Jena Axio Vert.A1) and a field-emission scanning electron microscope (FESEM, Nova Nano SEM450) were used to characterize the microstructural features. A backscattered electron (BSE) mode of FESEM was used to distinguish the reinforcements and steel matrix. The composition of the samples was analyzed using an energy-dispersive X-ray spectrometer (EDS) equipped on the FESEM. X-ray diffraction (XRD, D8 Advance) analyses with a Cu target were conducted for phase identification.

The microhardness of the samples was measured using a Vickers microhardness tester (Duramin-40, Struers, Denmark), with a 200 g load and a 10 s dwell time. Charpy U-notched impact tests were conducted with 55 mm × 10 mm × 10 mm samples on a pendulum impact machine (PTMS4300, Suns, China) at the RT. The notch was prepared perpendicular to the laser deposition direction. The reported impact toughness of each sample was averaged from three independent tests. The fracture surfaces of the impact samples were observed via FESEM.

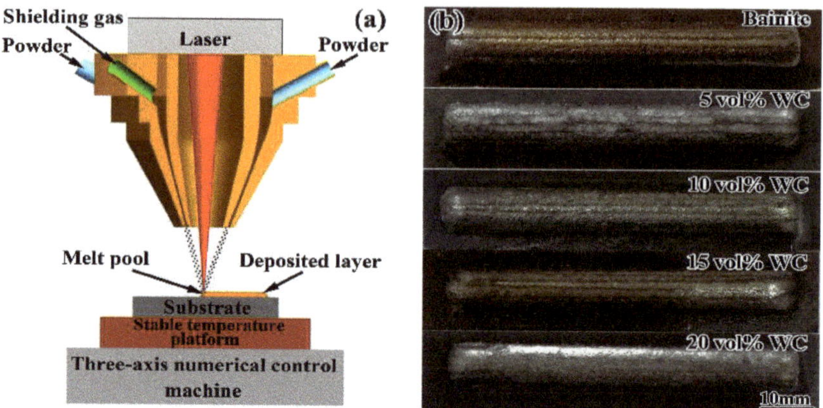

Figure 1. (**a**) Schematic of the laser processing system and (**b**) macroscopic morphology of the as-built composites.

3. Results

3.1. XRD Analysis

Figure 2 illustrates the phase constituents and their relative contents in the bainite steel matrix composite. As shown in Figure 2a, the matrix of the composite is changed from a mainly ferrite phase (α-Fe) with a little austenite phase (γ-Fe) to α-Fe with a considerable amount of γ-Fe when WC primary reinforcements are added. Furthermore, a Fe_3W_3C phase appears instead of a WC phase when the addition of primary reinforcement is relatively low. The WC phase presents together with the Fe_3W_3C phase when the primary reinforcement exceeds 15 vol%. The volume fraction of different phases was further quantified, as shown in Figure 2b. In the case of peak overlapping (the magnified images in Figure 2a), a Pearson VII function was used for the peak separation and fitting [15,40–42]. As the volume fraction of the WC primary reinforcement is increased, the volume fraction of the α-Fe phase declines; however, that of the γ-Fe phase increases. The volume fraction of the γ-Fe phase is not higher than that of the α-Fe phase until the WC primary reinforcement content reaches 20 vol%. As for the carbides, the volume fraction of Fe_3W_3C increases approximately linearly with the increased primary reinforcement content, and its maximum value is about 14 vol%. However, with the increased WC primary reinforcement content, the final WC content of the composites is maintained at zero when the content is less than 10%; meanwhile, when the content of primary reinforcement is higher than 10 vol%, the volume fraction of the final WC rises. When the content of primary reinforcement is 20%, the final WC reinforcement content in the composites is 4.1 vol%.

3.2. Microstructure

Figure 3 presents the optical micrographs of the bainite steel matrix composites with different volume fractions of WC. For the bainite steel without WC, the morphology of prior austenite grains can hardly be recognized. However, the prior austenite grains in the bainite steel matrix composites show a typical dendritic shape. Meanwhile, both the primary dendrite arm spacing (PDAS) and secondary dendrite arm spacing (SDAP) decrease with the increased volume fraction of WC. When the WC primary reinforcement content is higher than 15 vol%, plenty of white undissolved particles can be observed. The average diameter of the particles is about 47 μm, which is a little smaller than the average particle size of WC powder (about 65 μm). With the increased WC volume fraction, the bainite morphology changes from a lath shape to a needle shape and the content of block-like retained austenite (RA) increases, which is consistent with the XRD results (Figure 2b).

Furthermore, the black network-like microstructure begins to emerge in the interdendritic region when the WC primary reinforcement volume fraction is higher than 10%.

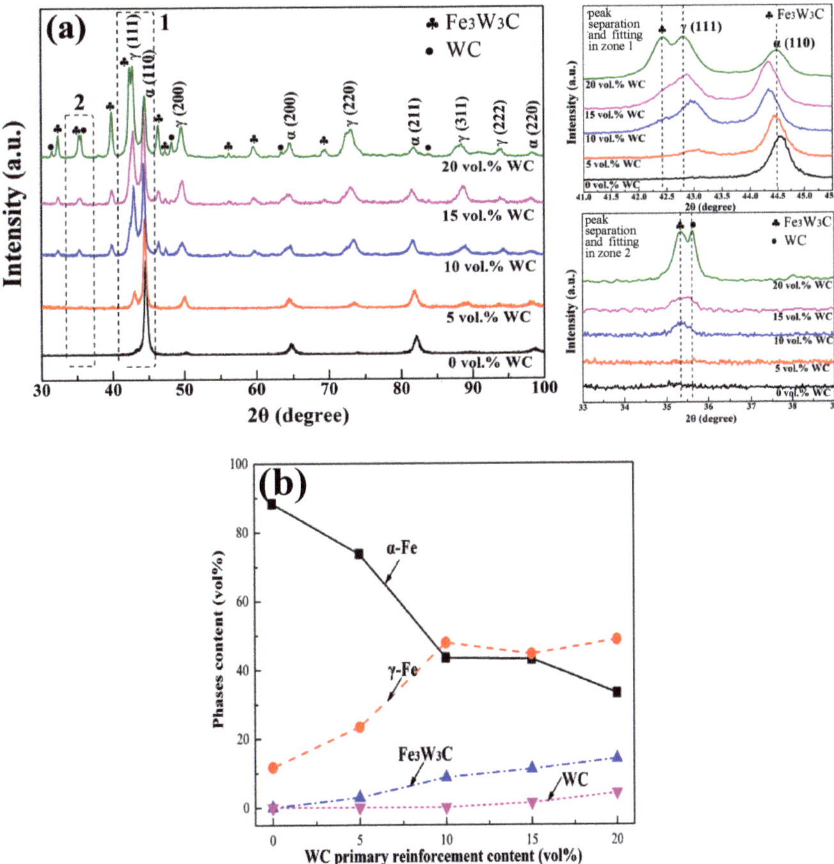

Figure 2. (a) XRD patterns and (b) the corresponding analyzed phase content in the bainite steel matrix composites.

The fine microstructure of the bainite steel matrix composites was further investigated using the BSE mode of SEM, as shown in Figure 4. As for the bainite steel with no WC addition (Figure 4a), the bainite steel mainly consists of sheaves of lath-like bainite, a few granular bainite (GB) and island-like RA. With the increased WC reinforcement volume fraction, the lath-like bainite and GB transform into black needle-like lower bainite (LB), and the length and width of the LB needles decrease gradually (Figure 4b–e). Meanwhile, the morphology of RA also changes from an island-like to block-like shape. For the bainite steel matrix composite with a relatively high volume fraction of WC (no less than 10 vol%), the white fish-bone-shaped microstructure appears at the boundary of the prior austenite grains (Figure 4d,e). With the increased addition of WC, the area of the intergranular region occupied by the white fish-bone-shaped microstructure increases; at the same time, the prior austenite grain is refined.

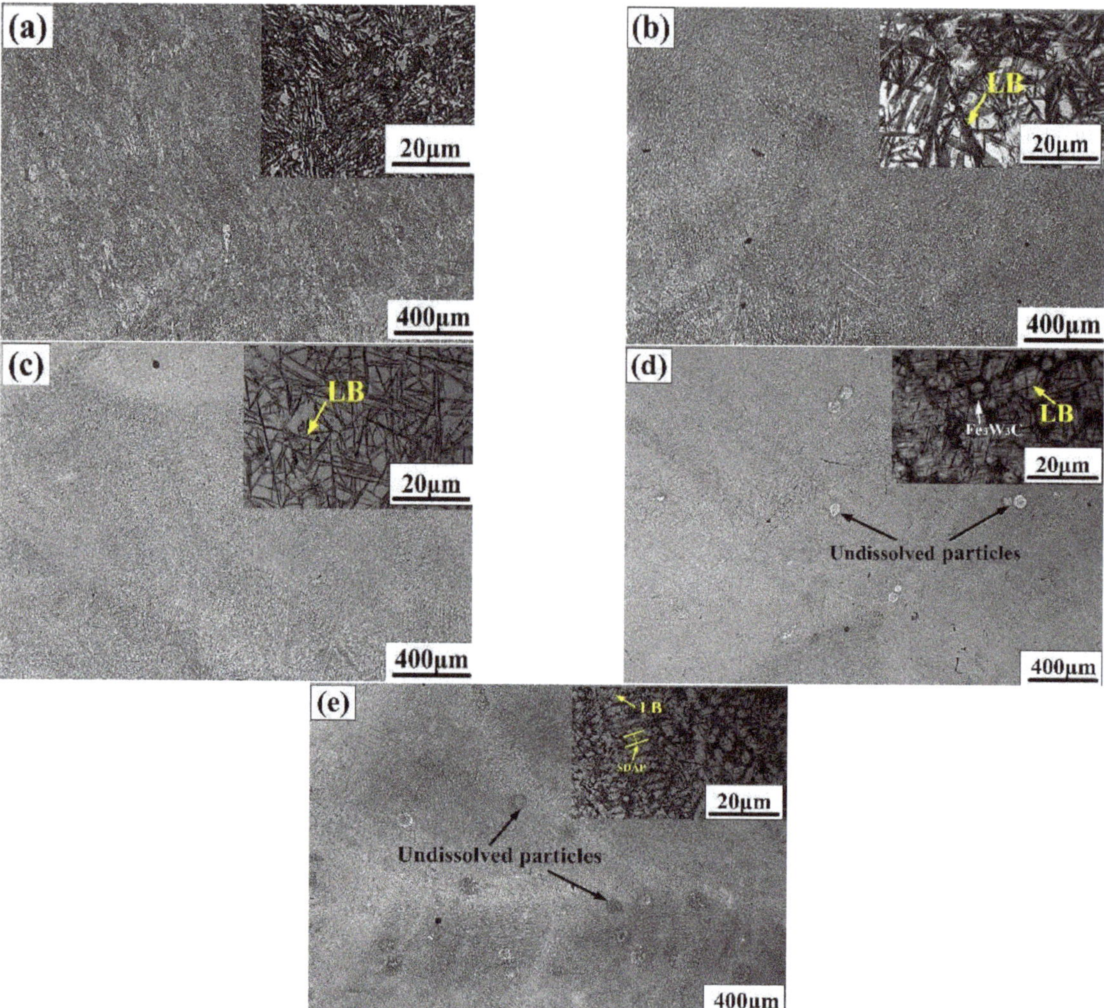

Figure 3. Microstructure evolution of the bainite steel matrix composites with the increased WC primary reinforcement volume fractions of (**a**) 0%, (**b**) 5%, (**c**) 10%, (**d**) 15%, and (**e**) 20%.

According to the phase constituent and the microstructure of the bainite steel matrix composite with 15 vol% WC particles, EDS analysis was conducted to further identify the phase composition of the intergranular region and undissolved particles. As shown in Figure 5a, the elemental maps of the intergranular region indicate that W enriches the white fish-bone-shaped microstructure. Combining the volume fraction of the white fish-bone-shaped microstructure obtained from the SEM images with the phase analysis results from the XRD patterns, the white fish-bone-shaped phase is Fe_3W_3C. The elemental distributions of the partial dissolved particles and the undissolved particles are presented in Figure 5b. Much W and little Fe can be detected in the partially dissolved WC particles region. Meanwhile, in the undissolved WC particles region, the enrichment degree of W in the interior of the particles is much higher than that of the partially dissolved particles, and no Fe is detected.

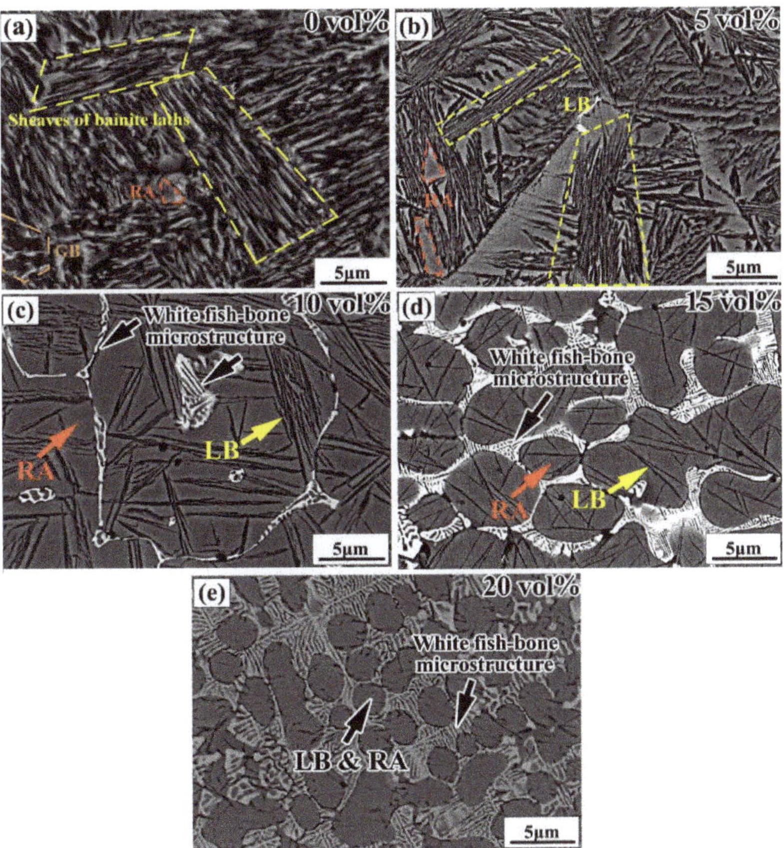

Figure 4. SEM images of the bainite steel matrix composites with the increased volume fraction of WC primary reinforcement (**a**) 0%, (**b**) 5%, (**c**) 10%, (**d**) 15%, and (**e**) 20%.

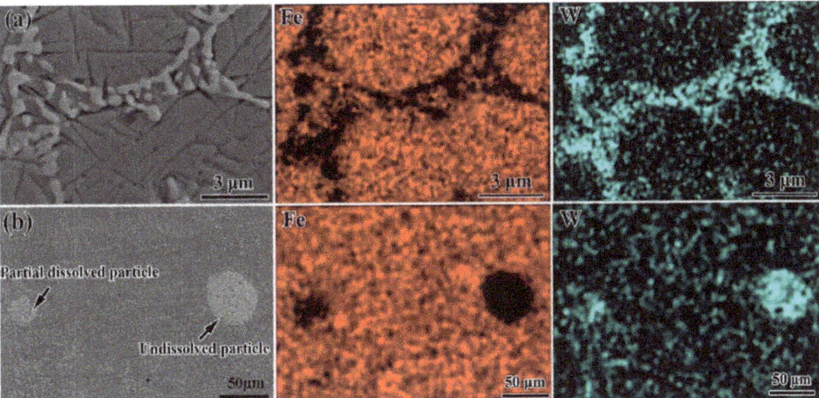

Figure 5. EDS mapping results at different regions: (**a**) intergranular region and (**b**) undissolved particles for the bainite steel matrix composite with 15 vol% WC primary reinforcement.

3.3. Mechanical Properties

As shown in Figure 6, the microhardness of the bainite steel is about 330 HV0.2, which is lower than that of the U75V steel substrate (375 HV0.2). The microhardness of the DLD manufactured bainite steel matrix composites is much higher than that of the U75V steel substrate. With the increased primary reinforcement content, the microhardness of bainite steel matrix composites increases remarkably. The microhardness increases rapidly to 461 HV0.2 when the WC primary reinforcement volume fraction is only 5 vol%, which is approximately 40% higher than that of the bainite steel. Moreover, the microhardness of composites with 20 vol% primary reinforcement is increased to 561 HV0.2. In contrast, the impact toughness of the bainite steel matrix composite decreases with the increased volume fraction of WC primary reinforcement (Figure 6). However, the impact toughness of the composite when the primary reinforcement volume fraction is less than 10% is still higher than that of the U75V steel substrate (26 J), which can satisfy the demand of the rail turnout service.

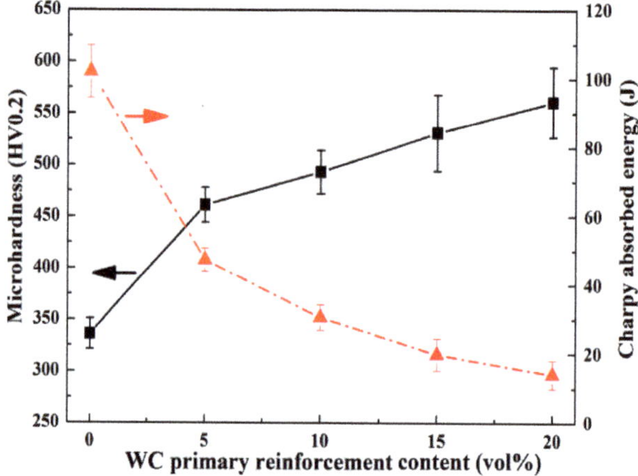

Figure 6. Microhardness and impact toughness of the bainite steel matrix composites.

Figure 7 indicates the impact fracture surface morphology of the U75V steel substrate and bainite steel matrix composites with different volume fractions of WC primary reinforcement. The U75V steel substrate shows a typical feature of cleavage fracture, which consists of cleavage steps and river patterns (Figure 7a). In contrast, dimples and tearing ridges are observed in the fracture of the bainite steel (Figure 7b), which indicates that the facture mechanism occurring is microvoids coalescence ductile fracture. Figure 7c presents the fracture of the bainite steel matrix composite with 5 vol% primary reinforcement. Both the features of ductile fracture (dimples and tearing ridges) and cleavage fracture (cleavage steps and river patterns) are evident on the fracture surface. This suggests that the fracture mechanism occurring is quasi-cleavage fracture in the composites. The crystal sugar fracture (Figure 7d) indicates that the bainite steel matrix composite with 15 vol% primary reinforcement follows the brittle intergranular fracture mechanism.

4. Discussion

4.1. Formation Mechanism of the Reinforcement during Laser Deposition

Information about the phases in the bainite steel matrix composite that was revealed by XRD (Figure 2) indicates that the in situ Fe_3W_3C reinforcement is the main reinforcement in the composite when the volume fraction of primary reinforcement WC is not higher than 10%. When the primary reinforcement volume fraction increases from 10% to 20%, the final reinforcement in the composites includes Fe_3W_3C in situ reinforcement and WC primary reinforcement. Moreover, the content of Fe_3W_3C is much higher than that of WC. Therefore, it can be deduced that the decomposed WC during laser deposition has dissolved in the bainite steel matrix and participated in the formation of Fe_3W_3C in the matrix. This is proved by the morphology of the Fe_3W_3C phase adjacent to the WC particles (Figure 8). The Fe_3W_3C phase in the vicinity of the partially dissolved WC particle presents the feature of a continuously fish-bone-shaped microstructure. The content of the Fe_3W_3C phase decreases with the increased distance from the partially dissolved WC particles. In contrast, compared with the Fe_3W_3C phase near the partially dissolved WC particle, the Fe_3W_3C phase near the undissolved WC particle has a much lower content and the carbides are distributed more uniformly.

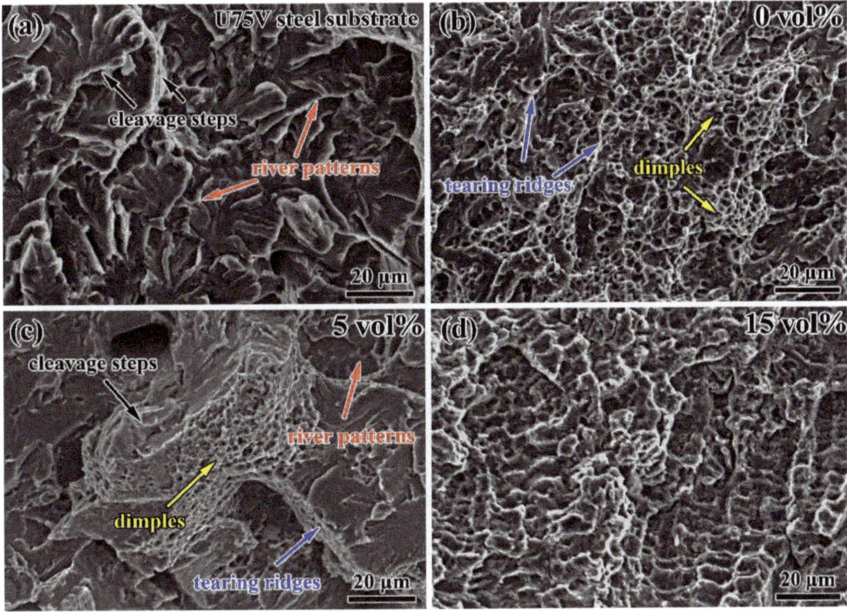

Figure 7. The impact fracture surface morphology of (**a**) the U75V steel substrate and (**b–d**) bainite steel matrix composites with different volume fractions of WC primary reinforcement.

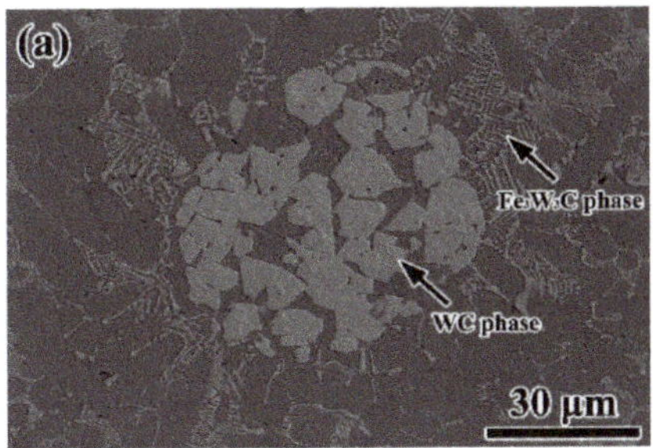

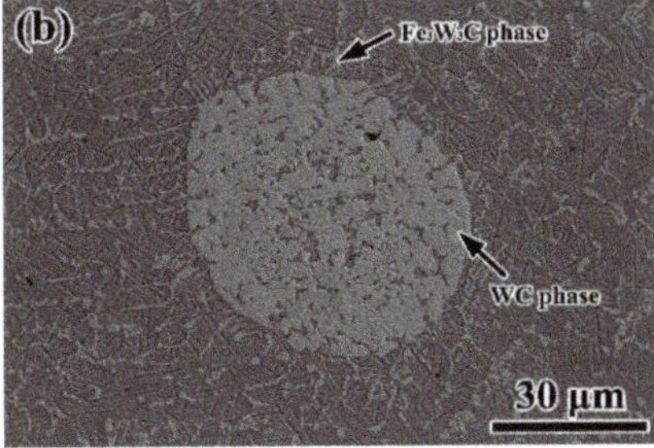

Figure 8. The morphology of the Fe_3W_3C adjacent to the (**a**) partially dissolved WC particle and (**b**) undissolved WC particle.

For the preparation of WC reinforced Fe-based matrix composites, the temperature of the material during direct laser deposition is always higher than that during traditional powder metallurgy and casting process [43–45]. Under the irradiation of a laser beam with an extremely high-energy density, the maximum temperature of the metal molten pool can exceed 3000 °C, which is higher than the decomposition temperature of WC (1250 °C) [46]. Then, the WC primary reinforcement will react with the steel matrix. The reaction is as follows [47,48]:

$$2WC \rightarrow W_2C + C \tag{1}$$

$$W_2C \rightarrow 2W + C \tag{2}$$

$$WC + W + L(\text{rich in Fe}) \rightarrow Fe_3W_3C\ (M_6C) \tag{3}$$

Additionally, the high temperature gradient of the molten pool during the DLD process brings the obvious convection of liquid metal, which is beneficial to a successful reaction [49]. When the addition of WC is low (less than 10 vol% in this work), all WC particles participate in the formation of Fe_3W_3C in the bainite steel matrix. In contrast, when WC is excessive, such as the 15 vol% WC in this work, both the Fe_3W_3C in situ reinforcement and WC primary reinforcement exist in the matrix at the same time. Free W atoms, which join to form Fe_3W_3C, come from the dissolved WC particle. Therefore, the concentration of W atoms around the partially dissolved WC particles is higher than that of the matrix far from the WC particles. Accordingly, the in situ Fe_3W_3C content near the partially dissolved WC particles is high.

4.2. Mechanism of the Adaptive Adjustment of the Matrix Microstructure with Primary Reinforcement

With the increased volume fraction of WC primary reinforcement, not only the constituent and morphology of the final reinforcement in the composites are changed, but also the microstructure of the bainite steel matrix is significantly altered (Figures 3 and 4), which is attributed to the change in the solute W and C content in the bainite steel matrix. Figure 9 presents the constituent of C and W derived from the WC primary reinforcement. It can be concluded that the formation of the Fe_3W_3C phase and WC phase does not consume all the C and W derived from the WC primary reinforcement and the remaining C and W dissolve in the bainite steel matrix, which significantly retards the transformation of undercooled austenite [50,51].

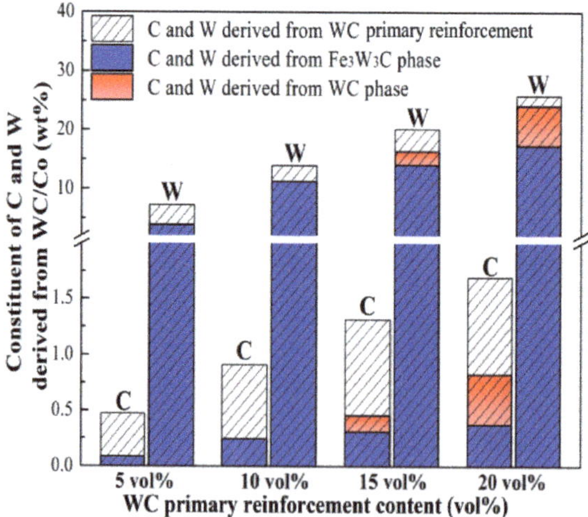

Figure 9. Constituent of C and W derived from primary reinforcement in the composites.

JMatPro software version 7.0 was applied to investigate the austenite isothermal transformation kinetics of the matrix of the composite using the general steel database (Figure 10). For the bainite steel (Figure 10a), the nose temperature of the pearlite transformation is 607 °C, with an incubation period of about 3 min; the nose temperature of the bainite transformation is 397 °C, with an incubation period of about 0.5 min. Meanwhile, the martensite transformation starts at 247 °C. Owing to the rapid cooling rate of the DLD process ($10^3 \sim 10^4$ °C/s), the pearlite transformation is completely prevented during the DLD process. In order to obtain the lower bainite microstructure, the transformation of martensite must be avoided the during cooling process and the following isothermal temperature has to be lower than the nose temperature of the bainite transformation. Hence, the preheating and isothermal temperature is set as 300 °C. According to the bainite transformation time at 300 °C, the isothermal treatment time was set to 200 min in this work.

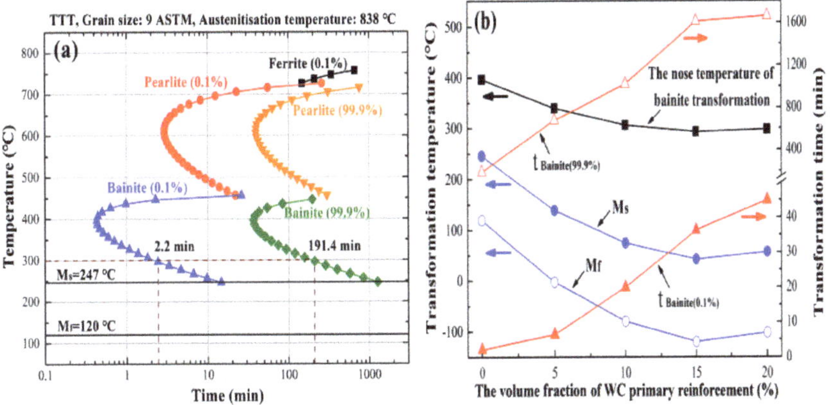

Figure 10. The austenite isothermal transformation kinetic of the composites: (**a**) time–temperature transformation (TTT) diagram of the bainite steel; (**b**) bainite and martensite transformation information of the composite with different volume fractions of WC primary reinforcement.

The nose temperature of the bainite transformation and the martensite transformation temperature of the bainite steel matrix in the composite were also calculated using JMatPro software, as well as the bainite transformation time (Figure 10b). With the increased volume fraction of primary reinforcement in the composite, both the nose temperature of bainite transformation and the martensite transformation temperature decrease. Furthermore, both the incubation period and completion time of the bainite transformation increase significantly when the primary reinforcement content in the composite is increased. The lowest bainite transformation nose temperature of the composite matrix is about 300 °C, which ensures that the bainite structure obtained via isothermal treatment in the matrix is needle-like lower bainite. When WC primary reinforcement is added into the bainite steel matrix, the start temperature of martensite transformation (M_s) in the matrix is still higher than room temperature, while the finish temperature of martensite transformation (M_f) in the matrix is much lower than room temperature. Meanwhile, the time required for the matrix to complete bainite transformation is much longer than the isothermal treatment time (200 min) for the composites. As a result, a small part of the untransformed undercooled austenite transforms into high-carbon cryptocrystalline martensite, while most of the untransformed undercooled austenite maintains its original crystal structure, which leads to the formation of a large amount of RA structure in the matrix. Moreover, the content of RA increases with the increased solute W and C content in the bainite steel matrix, which has a proportional relationship with the volume fraction of primary reinforcement in the composites (Figure 2b).

4.3. Effect of the Adaptive Adjustment of the Matrix Microstructure on the Mechanical Properties of Composites

Since reinforcement particles in MMCs are usually hard and brittle, the increased volume fraction of the reinforcement particles results in an enhancement in the strength and hardness of MMCs and a reduction in the ductility and toughness of MMCs (i.e., hardness-toughness trade-off). Figure 11 presents the hardness-toughness trade-off caused by the reinforcement particles' volume fraction in MMCs [52–59]. In order to facilitate a comparison between the mechanical properties of different MMCs, the microhardness and impact toughness are converted into the increase in microhardness and the decrease in impact toughness, respectively. Because the data point is close to the top right corner, the degree of hardness-toughness trade-off in the materials is low, and the hardness-toughness balance is good. The degree of hardness-toughness trade-off in the composite in this work is lower than that of most conventional MMCs, which can be attributed to the change in the matrix microstructure induced by the increased volume fraction of primary reinforcement. In conventional MMCs, the phase constituent and morphology of the matrix are almost not changed with the increased volume fraction of reinforcement, but the matrix microstructure of the composite in this work is altered with the increased primary reinforcement content. The residual austenite content in the steel matrix increases with the increased volume fraction of WC primary reinforcement, which is beneficial to improve the toughness of the steel matrix in composites [60]. The high volume fraction of reinforcement improves the hardness of the composite, while the adaptive adjustment of the matrix microstructure offsets a part of the reduction in impact toughness caused by the increase in the reinforcement volume fraction. As a result, the bainite steel matrix composite in this work demonstrates a better balance of hardness and impact toughness compared with most conventional MMCs.

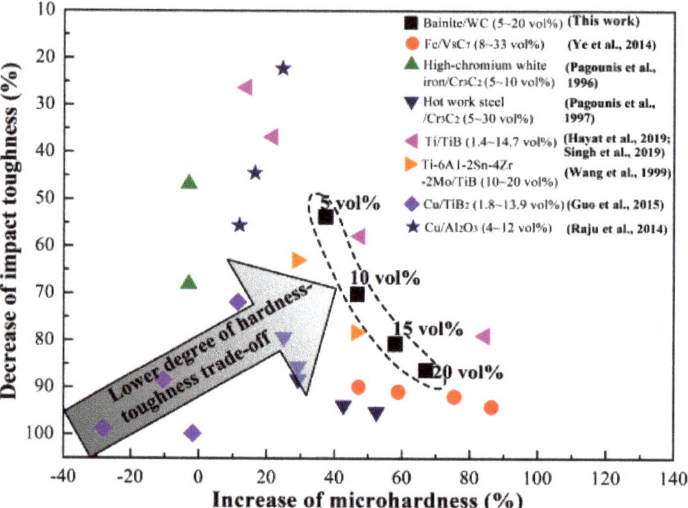

Figure 11. Hardness-toughness trade-off caused by the increased reinforcement volume fraction in MMCs (e.g., IMH = 100% × (HVBSC−20−HVBS)/HVBS, where IMH is the increase in microhardness, HVBSC−20 is the Vickers microhardness of the bainite steel matrix composite with 20 vol% primary reinforcement, and HVBS is the bainite steel) [52–59].

5. Conclusions

The in situ bainite steel matrix composites with WC primary reinforcement were manufactured using direct laser deposition. The effect of the primary reinforcement volume fraction on the composite microstructure and its mechanical properties were investigated. With the increased primary reinforcement content, the adaptive adjustment of the matrix microstructure was obtained. Furthermore, the dependence of the adaptive adjustment of the matrix on the combination of hardness and impact toughness in the composites was evaluated. The main conclusions are as follows:

(1) The interaction of the primary composite powder irradiated by laser during DLD leads to significant changes in the phase constituent and morphology of the reinforcement and matrix of the composites at the same time. With the increased volume fraction of primary reinforcement, the main phase in the matrix changes from dominant α-Fe to a mixture of γ-Fe and α-Fe. Specifically, with the increased primary reinforcement content, the matrix microstructure is changed from lath-like bainite, granular bainite and few island-like retained austenite into needle-like lower bainite and plenty of block-like retained austenite, and the final reinforcement is changed from Fe_3W_3C into Fe_3W_3C and WC.

(2) The microhardness increases rapidly to 461 HV0.2 when the WC primary reinforcement volume fraction is 5 vol%, which is approximately 40% higher than that of the bainite steel. Moreover, the microhardness of composites with 20 vol% primary reinforcement is increased to 561 HV0.2. In contrast, the impact toughness of the bainite steel matrix composite decreases with the increased primary reinforcement volume fraction. However, the impact toughness of the composite when the primary reinforcement volume fraction is less than 10% is still higher than that of the U75V steel substrate, which can satisfy the demand of rail turnout service.

(3) Compared with the conventional metal matrix composites, the bainite steel matrix composites manufactured via DLD possess a much lower degree of hardness-toughness trade-off. The better combination of microhardness and impact toughness can be attributed to the adaptive adjustment of the matrix microstructure with the increased volume fraction of primary reinforcement, which provides new insights into obtaining new materials with a good combination of hardness and toughness.

Author Contributions: Conceptualization, M.Z.; methodology, L.Z. (Longzhi Zhao) and H.Y.; validation, X.J. and Y.G.; formal analysis, M.Z. and Y.T.; investigation, D.L. and H.J.; resources, M.Z.; data curation, H.Y. and Y.G.; writing—original draft preparation, M.Z. and L.Z. (Longzhi Zhao); writing—review and editing, L.Z. (Laichang Zhang), L.Z. (Longzhi Zhao) and Y.T.; visualization, H.J.; supervision, M.Y.; project administration, L.Z. (Longzhi Zhao) and M.Y.; funding acquisition, M.Z. and L.Z. (Longzhi Zhao). All authors have read and agreed to the published version of the manuscript.

Funding: This research was funded from the National Natural Science Foundation of China (Grant Nos. 51965022, 51701074 and 51761012), Natural Science Foundation of Jiangxi Province (Grant Nos. 20202BABL204046 and 20192BAB206028) and Natural Science Foundation of Jiangxi Province Education Department (Grant Nos.GJJ 200606).

Data Availability Statement: Research data are not shared.

Acknowledgments: The authors would like to express our sincere thanks to Chengyu Guo (School of Materials Science and Engineering, University of Science and Technology Beijing) for his assistance in performing the calculation by using JMatPro software.

Conflicts of Interest: The authors declare no conflict of interest.

References

1. Nielsen, J.C.O.; Pålsson, B.A.; Torstensson, P.T. Switch panel design based on simulation of accumulated rail damage in a railway turnout. *Wear* **2016**, *366–367*, 241–248. [CrossRef]
2. Wang, W.J.; Guo, H.M.; Du, X.; Guo, J.; Liu, Q.Y.; Zhu, M.H. Investigation on the damage mechanism and prevention of heavy-haul railway rail. *Eng. Fail. Anal.* **2013**, *35*, 206–218. [CrossRef]
3. Liu, J.P.; Zhou, Q.Y.; Zhang, Y.H.; Liu, F.S.; Tian, C.H.; Li, C.; Zhi, X.Y.; Li, C.G.; Shi, T. The formation of martensite during the propagation of fatigue cracks in pearlitic rail steel. *Mater. Sci. Eng. A* **2019**, *747*, 199–205. [CrossRef]
4. Xin, L.; Markine, V.L.; Shevtsov, I.Y. Numerical procedure for fatigue life prediction for railway turnout crossings using explicit finite element approach. *Wear* **2016**, *366–367*, 167–179. [CrossRef]
5. Wang, P.; Xu, J.; Xie, K.; Chen, R. Numerical simulation of rail profiles evolution in the switch panel of a railway turnout. *Wear* **2016**, *366–367*, 105–115. [CrossRef]
6. Shu, F.Y.; Tian, Z.; Lü, Y.H.; He, W.X.; Lü, F.Y.; Lin, J.J.; Zhao, H.Y.; Xu, B.S. Prediction of vulnerable zones based on residual stress and microstructure in CMT welded aluminum alloy joint. *Trans. Nonferrous Met. Soc. China* **2015**, *25*, 2701–2707. [CrossRef]
7. Herzog, D.; Seyda, V.; Wycisk, E.; Emmelmann, C. Additive manufacturing of metals. *Acta Mater.* **2016**, *117*, 371–392. [CrossRef]
8. DebRoy, T.; Wei, H.L.; Zuback, J.S.; Mukherjee, T.; Elmer, J.W.; Milewski, J.O.; Beese, A.M.; Wilson-Heid, A.; De, A.; Zhang, W. Additive manufacturing of metallic components—Process, structure and properties. *Prog. Mater. Sci.* **2018**, *92*, 112–224. [CrossRef]

9. Lai, Q.; Abrahams, R.; Yan, W.; Qiu, C.; Mutton, P.; Paradowska, A.; Soodi, M. Investigation of a novel functionally graded material for the repair of premium hypereutectoid rails using laser cladding technology. *Compos. Part B Eng.* **2017**, *130*, 174–191. [CrossRef]
10. Lai, Q.; Abrahams, R.; Yan, W.; Qiu, C.; Mutton, P.; Paradowska, A.; Fang, X.; Soodi, M.; Wu, X. Effects of preheating and carbon dilution on material characteristics of laser-cladded hypereutectoid rail steels. *Mater. Sci. Eng. A* **2018**, *712*, 548–563. [CrossRef]
11. Guo, Y.; Feng, K.; Lu, F.; Zhang, K.; Li, Z.; Hosseini, S.R.E.; Wang, M. Effects of isothermal heat treatment on nanostructured bainite morphology and microstructures in laser cladded coatings. *Appl. Surf. Sci.* **2015**, *357*, 309–316. [CrossRef]
12. Xing, X.L.; Zhou, Y.F.; Yang, Y.L.; Gao, S.Y.; Ren, X.J.; Yang, Q.X. Surface modification of low-carbon nano-crystallite bainite via laser remelting and following isothermal transformation. *Appl. Surf. Sci.* **2015**, *353*, 184–188. [CrossRef]
13. Xing, X.L.; Yuan, X.M.; Zhou, Y.F.; Qi, X.W.; Lu, X.; Xing, T.H.; Ren, X.J.; Yang, Q.X. Effect of bainite layer by LSMCIT on wear resistance of medium-carbon bainite steel at different temperatures. *Surf. Coat. Technol.* **2017**, *325*, 462–472. [CrossRef]
14. Xing, X.L.; Zhou, Y.F.; Gao, S.Y.; Wang, J.B.; Yang, Y.L.; Yang, Q.X. Nano-twin in surface modified bainite induced by laser surface modification. *Mater. Lett.* **2016**, *165*, 79–82. [CrossRef]
15. Guo, Y.; Li, Z.; Yao, C.; Zhang, K.; Lu, F.; Feng, K.; Huang, J.; Wang, M.; Wu, Y. Microstructure evolution of Fe-based nanostructured bainite coating by laser cladding. *Mater. Des.* **2014**, *63*, 100–108. [CrossRef]
16. Miracle, D.B. Metal matrix composites–from science to technological significance. *Compos. Sci. Technol.* **2005**, *65*, 2526–2540. [CrossRef]
17. Bhoi, N.K.; Singh, H.; Pratap, S. Developments in the aluminum metal matrix composites reinforced by micro/nano particles—A review. *J. Compos. Mater.* **2020**, *54*, 813–833. [CrossRef]
18. Kim, C.S.; Cho, K.; Manjili, M.H.; Nezafati, M. Mechanical performance of particulate-reinforced Al metal-matrix composites (MMCs) and Al metal-matrix nano-composites (MMNCs). *J. Mater. Sci.* **2017**, *52*, 13319–13349. [CrossRef]
19. Ferguson, J.B.; Lopez, H.F.; Rohatgi, P.K.; Cho, K.; Kim, C.S. Impact of volume fraction and size of reinforcement particles on the grain size in metal-matrix micro and nanocomposites. *Metall. Mater. Trans. A* **2014**, *45*, 4055–4061. [CrossRef]
20. Liu, J.; Yang, S.; Xia, W.; Jiang, X.; Gui, C. Microstructure and wear resistance performance of Cu-Ni-Mn alloy based hardfacing coatings reinforced by WC particles. *J. Alloys Compd.* **2016**, *654*, 63–70. [CrossRef]
21. Kenesei, P.; Biermann, H.; Borbély, A. Structure-property relationship in particle reinforced metal-matrix composites based on holotomography. *Scr. Mater.* **2005**, *53*, 787–791. [CrossRef]
22. Yang, X.; Wu, G.Q.; Sha, W.; Zhang, Q.Q.; Huang, Z. Numerical study of the effects of reinforcement/matrix interphase on stress-strain behavior of YAl2 particle reinforced MgLiAl composites. *Compos. Part A* **2012**, *43*, 363–369. [CrossRef]
23. Smirnov, S.V.; Konovalov, A.V.; Myasnikova, M.V.; Khalevitsky, Y.V.; Smirnov, A.S.; Igumnov, A.S. A numerical study of plastic strain localization and fracture in Al/SiC metal matrix composite. *Phys. Mesomech.* **2018**, *21*, 305–313. [CrossRef]
24. Zhang, R.; Wang, D.J.; Huang, L.J.; Yuan, S.J.; Geng, L. Deformation behaviors and microstructure evolution of TiBw/TA15 composite with novel network architecture. *J. Alloys Compd.* **2017**, *722*, 970–980. [CrossRef]
25. Liang, S.; Li, W.; Jiang, Y.; Cao, F.; Dong, G.; Xiao, P. Microstructures and properties of hybrid copper matrix composites reinforced by TiB whiskers and TiB_2 particles. *J. Alloys Compd.* **2019**, *797*, 589–594. [CrossRef]
26. Peng, H.X.; Fan, Z.; Evans, J.R.G. Bi-continuous metal matrix composites. *Mater. Sci. Eng. A* **2001**, *303*, 37–45. [CrossRef]
27. Rodríguez-Castro, R.; Wetherhold, R.C.; Kelestemur, M.H. Microstructure and mechanical behavior of functionally graded Al A359/SiC_p composite. *Mater. Sci. Eng. A* **2002**, *323*, 445–456. [CrossRef]
28. Almangour, B.; Grzesiak, D.; Yang, J.M. Rapid fabrication of bulk-form TiB_2/316L stainless steel nanocomposites with novel reinforcement architecture and improved performance by selective laser melting. *J. Alloys Compd.* **2016**, *680*, 480–493. [CrossRef]
29. AlMangour, B.; Grzesiak, D.; Yang, J.M. Selective laser melting of TiC reinforced 316L stainless steel matrix nanocomposites: Influence of starting TiC particle size and volume content. *Mater. Des.* **2016**, *104*, 141–151. [CrossRef]
30. AlMangour, B.; Grzesiak, D.; Yang, J.M. Selective laser melting of TiB_2/316L stainless steel composites: The roles of powder preparation and hot isostatic pressing post-treatment. *Powder Technol.* **2017**, *309*, 37–48. [CrossRef]
31. AlMangour, B.; Baek, M.S.; Grzesiak, D.; Lee, K.A. Strengthening of stainless steel by titanium carbide addition and grain refinement during selective laser melting. *Mater. Sci. Eng. A* **2018**, *712*, 812–818. [CrossRef]
32. Attar, H.; Bönisch, M.; Calin, M.; Zhang, L.C.; Scudino, S.; Eckert, J. Selective laser melting of in situ titanium-titanium boride composites: Processing, microstructure and mechanical properties. *Acta Mater.* **2014**, *76*, 13–22. [CrossRef]
33. Rong, T.; Gu, D.; Shi, Q.; Cao, S.; Xia, M. Effects of tailored gradient interface on wear properties of WC/Inconel 718 composites using selective laser melting. *Surf. Coat. Technol.* **2016**, *307*, 418–427. [CrossRef]
34. Muvvala, G.; Patra Karmakar, D.; Nath, A.K. Online assessment of TiC decomposition in laser cladding of metal matrix composite coating. *Mater. Des.* **2017**, *121*, 310–320. [CrossRef]
35. Tang, Y.; Yang, H.; Huang, D.; Zhao, L.; Liu, D.; Shen, M.; Hu, Y.; Zhao, M.; Zhang, J.; Li, J.; et al. Dual-gradient bainite steel matrix composite fabricated by direct laser deposition. *Mater. Lett.* **2019**, *238*, 210–213. [CrossRef]
36. Hu, Y.; Cong, W. A review on laser deposition-additive manufacturing of ceramics and ceramic reinforced metal matrix composites. *Ceram. Int.* **2018**, *44*, 20599–20612. [CrossRef]
37. Muvvala, G.; Patra Karmakar, D.; Nath, A.K. In-process detection of microstructural changes in laser cladding of in-situ Inconel 718/TiC metal matrix composite coating. *J. Alloys Compd.* **2018**, *740*, 545–558. [CrossRef]
38. Przybylowicz, J.; Kusiński, J. Structure of laser cladded tungsten carbide composite coatings. *J. Mater. Process. Technol.* **2001**, *109*, 154–160. [CrossRef]

39. Wang, Z.; Wang, J.; Dong, H.; Zhou, Y.; Jiang, F. Hardening through an ultrafine carbide precipitation in austenite of a low-carbon steel containing titanium and tungsten. *Metall. Mater. Trans. A* **2020**, *51*, 3778–3788. [CrossRef]

40. De, A.K.; Murdock, D.C.; Mataya, M.C.; Speer, J.G.; Matlock, D.K. Quantitative measurement of deformation-induced martensite in 304 stainless steel by X-ray diffraction. *Scr. Mater.* **2004**, *50*, 1445–1449. [CrossRef]

41. Zhang, L.C.; Shen, Z.Q.; Xu, J. Glass formation in a (Ti,Zr,Hf)-(Cu,Ni,Ag)-Al high-order alloy system by mechanical alloying. *J. Mater. Res.* **2003**, *18*, 2141–2149. [CrossRef]

42. Rabadia, C.D.; Liu, Y.J.; Jawed, S.F.; Wang, L.; Li, Y.H.; Zhang, X.H.; Sercombe, T.B.; Sun, H.; Zhang, L.C. Improved deformation behavior in Ti-Zr-Fe-Mn alloys comprising the C14 type Laves and β phases. *Mater. Des.* **2018**, *160*, 1059–1070. [CrossRef]

43. Hashim, J.; Looney, L.; Hashmi, M.S.J. Metal matrix composites: Production by the stir casting method. *J. Mater. Process. Technol.* **1999**, *92–93*, 1–7. [CrossRef]

44. Kaczmar, J.W.; Pietrzak, K.; Wlosiński, W. Production and application of metal matrix composite materials. *J. Mater. Process. Technol.* **2000**, *106*, 58–67. [CrossRef]

45. Yu, W.H.; Sing, S.L.; Chua, C.K.; Kuo, C.N.; Tian, X.L. Particle-reinforced metal matrix nanocomposites fabricated by selective laser melting: A state of the art review. *Prog. Mater. Sci.* **2019**, *104*, 330–379. [CrossRef]

46. Zhang, D.Y.; Feng, Z.; Wang, C.J.; Liu, Z.; Dong, D.D.; Zhou, Y.; Wu, R. Modeling of temperature field evolution during multilayered direct laser metal deposition. *J. Therm. Spray Technol.* **2017**, *26*, 831–845. [CrossRef]

47. Yang, J.; Miao, X.; Wang, X.; Yang, F. Influence of Mn additions on the microstructure and magnetic properties of FeNiCr/60% WC composite coating produced by laser cladding. *Int. J. Refract. Met. Hard Mater.* **2014**, *46*, 58–64. [CrossRef]

48. Zhou, S.; Lei, J.; Dai, X.; Guo, J.; Gu, Z.; Pan, H. A comparative study of the structure and wear resistance of NiCrBSi/50 wt.% WC composite coatings by laser cladding and laser induction hybrid cladding. *Int. J. Refract. Met. Hard Mater.* **2016**, *60*, 17–27. [CrossRef]

49. Lee, Y.; Nordin, M.; Babu, S.S.; Farson, D.F. Effect of fluid convection on dendrite arm spacing in laser deposition. *Metall. Mater. Trans. B* **2014**, *45*, 1520–1529. [CrossRef]

50. Tong, X.; Li, F.H.; Kuang, M.; Ma, W.Y.; Chen, X.C.; Liu, M. Effects of WC particle size on the wear resistance of laser surface alloyed medium carbon steel. *Appl. Surf. Sci.* **2012**, *258*, 3214–3220. [CrossRef]

51. Ghosh, L.; Ma, L.; Ofori-Opoku, N.; Guyer, J.E. On the primary spacing and microsegregation of cellular dendrites in laser deposited Ni-Nb alloys. *Model. Simul. Mater. Sci. Eng.* **2017**, *25*, 065002. [CrossRef]

52. Ye, F.; Hojamberdiev, M.; Xu, Y.; Zhong, L.; Yan, H.; Chen, Z. Volume fraction effect of V8C7 particulates on impact toughness and wear performance of V_8C_7/Fe monolithic composites. *J. Mater. Eng. Perform.* **2014**, *23*, 1402–1407. [CrossRef]

53. Pagounis, E.; Talvitie, M.; Lindroos, V.K. Influence of the metal/ceramic interface on the microstructure and mechanical properties of HIPed iron-based composites. *Compos. Sci. Technol.* **1996**, *56*, 1329–1337. [CrossRef]

54. Pagounis, E.; Lindroos, V.K. Development and performance of new hard and wear-resistant engineering materials. *J. Mater. Eng. Perform.* **1997**, *6*, 749–756. [CrossRef]

55. Hayat, M.D.; Singh, H.; He, Z.; Cao, P. Titanium metal matrix composites: An overview. *Compos. Part A* **2019**, *121*, 418–438. [CrossRef]

56. Singh, H.; Hayat, M.; Zhang, H.Z.; Das, R.; Cao, P. Effect of TiB_2 content on microstructure and properties of in situ Ti–TiB composites. *Int. J. Miner. Metall. Mater.* **2019**, *26*, 915–924. [CrossRef]

57. Wang, L.; Niinomi, M.; Takahashi, S.; Hagiwara, M.; Emura, S.; Kawabei, Y.; Sung-Joon, K. Relationship between fracture toughness and microstructure of Ti-6Al-2Sn-4Zr-2Mo alloy reinforced with TiB particles. *Mater. Sci. Eng. A* **1999**, *263*, 319–325. [CrossRef]

58. Guo, Z.; Li, N.; Hu, J. Cu-TiB metal matrix composites prepared by powder metallurgy route. *Sci. Sinter.* **2015**, *47*, 165–174. [CrossRef]

59. Raju, L.S.; Kumar, A. A novel approach for fabrication of Cu-Al2O3 surface composites by friction stir processing. *Procedia Mater. Sci.* **2014**, *5*, 434–443. [CrossRef]

60. Wu, S.; Wang, D.; Di, X.; Li, C.; Zhang, Z.; Zhou, Z.; Liu, X. Strength-toughness improvement of martensite-austenite dual phase deposited metals after austenite reversed treatment with short holding time. *Mater. Sci. Eng. A* **2019**, *755*, 57–65. [CrossRef]

materials

Article

Effect of Porosity and Heat Treatment on Mechanical Properties of Additive Manufactured CoCrMo Alloys

Tu-Ngoc Lam [1,2], Kuang-Ming Chen [1], Cheng-Hao Tsai [1], Pei-I Tsai [3], Meng-Huang Wu [4,5], Ching-Chi Hsu [6], Jayant Jain [7] and E-Wen Huang [1,*]

1   Department of Materials Science and Engineering, National Yang Ming Chiao Tung University, Hsinchu 30010, Taiwan; lamtungoc1310@nycu.edu.tw (T.-N.L.); chenkevin24@gmail.com (K.-M.C.); za95139@gmail.com (C.-H.T.)
2   Department of Physics, College of Education, Can Tho University, Can Tho City 900000, Vietnam
3   Biomedical Technology and Device Research Laboratories, Industrial Technology Research Institute, Chutung, Hsinchu 310401, Taiwan; peiyi@itri.org.tw
4   Department of Orthopaedics, Taipei Medical University Hospital, Taipei 11031, Taiwan; maxwutmu@gmail.com
5   Department of Orthopedics, College of Medicine, Taipei Medical University, No. 250, Wuxing St., Xinyi District, Taipei 11031, Taiwan
6   Department of Mechanical Engineering, National Taiwan University of Science and Technology, Taipei 106, Taiwan; hsucc@mail.ntust.edu.tw
7   Department of Materials Science and Engineering, Indian Institute of Technology, New Delhi 110016, India; jayant.jain@mse.iitd.ac.in
*   Correspondence: ewenhuang@nycu.edu.tw

Abstract: To minimize the stress shielding effect of metallic biomaterials in mimicking bone, the body-centered cubic (bcc) unit cell-based porous CoCrMo alloys with different, designed volume porosities of 20, 40, 60, and 80% were produced via a selective laser melting (SLM) process. A heat treatment process consisting of solution annealing and aging was applied to increase the volume fraction of an ε-hexagonal close-packed (hcp) structure for better mechanical response and stability. In the present study, we investigated the impact of different, designed volume porosities on the compressive mechanical properties in as-built and heat-treated CoCrMo alloys. The elastic modulus and yield strength in both conditions were dramatically decreased with increasing designed volume porosity. The elastic modulus and yield strength of the CoCrMo alloys with a designed volume porosity of 80% exhibited the closest match to those of bone tissue. Different strengthening mechanisms were quantified to determine their contributing roles to the measured yield strength in both conditions. The experimental results of the relative elastic modulus and yield strength were compared to the analytical and simulation modeling analyses. The Gibson–Ashby theoretical model was established to predict the deformation behaviors of the lattice CoCrMo structures.

Keywords: selective laser melting; CoCrMo; porosity; heat treatment; compression test

check for **updates**

Citation: Lam, T.-N.; Chen, K.-M.; Tsai, C.-H.; Tsai, P.-I.; Wu, M.-H.; Hsu, C.-C.; Jain, J.; Huang, E.-W. Effect of Porosity and Heat Treatment on Mechanical Properties of Additive Manufactured CoCrMo Alloys. *Materials* **2023**, *16*, 751. https://doi.org/10.3390/ma16020751

Academic Editor: Federico Mazzucato

Received: 12 December 2022
Revised: 4 January 2023
Accepted: 9 January 2023
Published: 12 January 2023

1. Introduction

Metallic materials, such as tantalum (Ta)-based, titanium (Ti)-based, and cobalt (Co)-based alloys, are widely used as bone implants. Among the most prevalent materials used for promising biomedical applications [1], cobalt-chromium-molybdenum (CoCrMo) alloys [2] have attracted great interest due to their superior biocompatibility, corrosion resistance, wear resistance, and good mechanical properties [3–10]. Poor tribological behavior caused by a high friction coefficient and wear debris is one of the crucial obstacles for Ti-based alloys [11], which can be overcome by the use of CoCrMo alloys. Extensive research on the mechanical and microstructural properties of widely used cast, wrought, or hot-forged CoCrMo alloys has been reported [6,12–15]. To fulfill the criteria for metallic materials as feasible implants, it is important to reduce the stress shielding effect and enhance osseointegration [16–21] by applying porous lattice structures, which are difficult to achieve via traditional fabrication processes.

The wear resistance of CoCrMo alloys is governed by the amount of carbon, the homogeneous distribution of carbides, and the existence of a hexagonal close-packed (hcp) structure [22]. There may exist two different crystal structures—metastable γ-face-centered cubic (fcc) and ϵ-hcp—in the CoCrMo alloys at room temperature, and their volume fraction can be changed via heat treatment conditions [23–25]. Increasing the volume fraction of the hcp phase is beneficial to the improved mechanical and wear properties of CoCrMo alloys, as well as to their stability [24–28]. A number of studies on various process- and heat-treatment conditions have been devoted to promoting a martensitic transformation from fcc to hcp [23–26,29,30].

Additive manufacturing allows for complex structures produced with diverse geometries and shapes associated with controllable microstructures [31–35]. Better metallurgical design can be achieved through machine learning and high-throughput examinations [36]. Among the most common three-dimensional (3D) printing processes, selective laser melting (SLM) produces SLM-built metallic parts with distinct microstructures [37–39]. In addition, SLM offers great possibilities in tailoring porous lattice structures with various unit cell types, cell sizes, and strut dimensions, which enables the tuning of the mechanical properties of SLM-built metallic implants to closely match those of human bone [39,40]. Comprehensive research has been conducted on the mechanical properties of additive-manufactured porous CoCrMo alloys via different, designed volume porosities [41–43] or heat treatment conditions to reduce the stiffness mismatch between bone and biomedical CoCrMo implants [24,25,44]. Furthermore, the analytical and simulation modeling analyses were generally used to predict the mechanical properties of porous structures [18,45,46]. However, investigating the ideal porosity and pore sizes for porous implants is still controversial [47], and exploring the role of different, designed volume porosities on heat-treated CoCrMo alloys fabricated via SLM is limited.

In the present study, the designed lattice structure of body-centered cubic (bcc) unit cell-based porous CoCrMo alloys was manufactured via SLM. The objective of this work was to discover the optimal design parameters for a closer match between bone tissue and CoCrMo alloys. The influence of different, designed volume porosities on the compressive mechanical properties of as-built and heat-treated CoCrMo alloys was examined. Moreover, the theoretical model proposed by Gibson–Ashby was employed to predict the mechanical behavior of porous SLM-built CoCrMo structures, which is conducive to establishing the optimal design of bcc lattice structures with suitable mechanical properties for biomedical applications.

2. Materials and Methods

2.1. Sample Preparation

The cylindrical shapes of as-built porous CoCrMo alloys with a diameter of 11 mm and a height of 7 mm were fabricated using the SLM AM100 machine with a working space of 10 cm × 10 cm manufactured by the Industrial Technology Research Institute (ITRI). The diameters of the struts were 0.2, 0.3, 0.4, and 0.5 mm, corresponding to the designed volume porosity of 80, 60, 40, and 20%, respectively. Figure 1a–e shows computer aided design (CAD) models for the designed porous CoCrMo structures with different volume porosities of bcc unit cells with a length of 1 mm. The building direction was parallel to the longitudinal axis of the as-built CoCrMo samples. The chemical composition of fully dense as-built CoCrMo alloy was Co (58 wt%), Cr (28 wt%), Mo (6 wt%), and Si (<1 wt%).

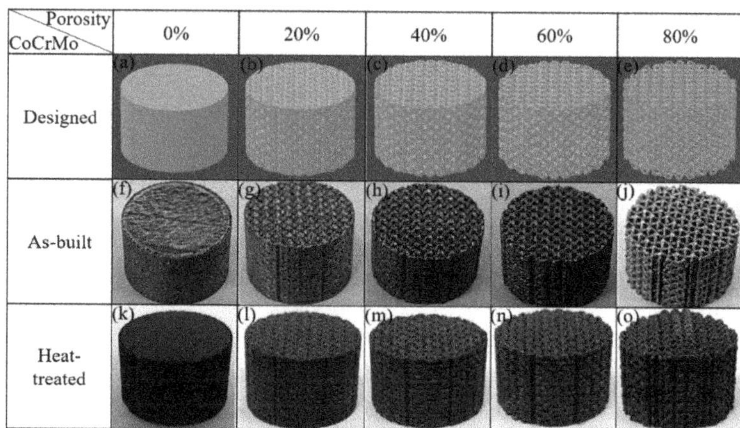

Figure 1. (**a–e**) CAD-designed models, (**f–j**) as-built, and (**k–o**) heat-treated CoCrMo alloys with different, designed volume porosities.

2.2. Heat Treatment Process

The heat-treated CoCrMo structures were prepared using a combination of solution heat treatment at 1100 °C for 1 h and subsequent aging treatment at 800 °C for 4 h of the as-built CoCrMo alloys.

2.3. Mechanical Test

The uniaxial compression tests of as-built and heat-treated CoCrMo alloys were carried out using an HT-2402 universal testing machine produced by the Hung Ta Company, Taichung, Taiwan, with a 50 kN load cell and a strain rate of 2.1×10^{-3} s^{-1} at room temperature. The samples used for mechanical tests with dimensions of 5 mm × 5 mm × 4 mm were cut from the CoCrMo alloys. The compression direction was parallel to the building direction.

2.4. Microstructure Characterization

The CoCrMo alloys were mechanically polished using silicon carbide sandpapers of 4000-grit and finally using 0.02 μm colloidal silica suspension. The samples were subsequently etched for microstructure characterization using optical microscope (OM, Nikon ECLIPSE LV150N, Minato ku, Japan) and scanning electron microscopy (SEM, JEOL 6700F, Akishima, Japan).

2.5. Density Measurement

The density of the foam is shown below,

$$\rho_f = \frac{M_{porous}}{V_{porous}} \tag{1}$$

where M_{porous} and V_{porous} are the weight and volume of the porous sample.

Meanwhile, the density of solid structure (ρ_S) was measured using the Archimedes method as follows,

$$\rho_S = \frac{\rho_w \times w_a}{w_a - w_w} \tag{2}$$

where ρ_w is the density of water, w_a and w_w are the weights of the sample in air and in water, respectively.

The relative density of material (ρ^*) was determined as

$$\rho^* = \frac{\rho_{measured}}{\rho_{theoretical}} \tag{3}$$

where $\rho_{measured}$ and $\rho_{theoretical}$ are the measured density, and the theoretical bulk density, $\rho_{theoretical}$, is 8.3 g/cm^3.

2.6. Finite Element Simulation

A finite element method using the ANSYS Explicit Dynamics (2019R1) software was simulated to predict the compressive mechanical behavior of the as-built and heat-treated porous CoCrMo with various designed porosities in comparison with the analytical prediction and experimental data. The 3D models with a length of 5 mm, a width of 5 mm, and a height of 4 mm exported from the 3D builder CAD software (18.0.1931.0), in STL file formats of the 3D-printed models were used for simulation to satisfy the actual compression test. The top and bottom plates were set as rigid bodies, while the 3D models were set to be deformable. The friction coefficient was set to 0.2, and the self-contact of 3D models was set. The Cartesian and tetrahedral mesh was used with a minimum element size of 0.00011 m, and the total number of the elements and nodes was around 220,000 and 59,000, respectively. The bottom plate had no freedom of x, y, and z displacements, and any rotation was restricted. The top plate was set with a z displacement while x and y displacements were restrained. The elastic and plastic behaviors were simulated based on an isotropic elasticity model and a bilinear isotropic hardening model, respectively. The values of Young's modulus, yield strength, and tangential modulus derived from the experimental data of the fully dense as-built and heat-treated CoCrMo alloys were input into the simulation analysis. A Poisson's ratio of 0.3 was applied [7]. A maximum plastic strain of 0.2 was set as the failure criterion of materials. If the plastic deformation of the element exceeded 0.2, it was considered damaged and thus removed.

3. Results

3.1. Morphology of As-Built and Heat-Treated Porous CoCrMo Alloys

Figure 1f–o shows the cylinder shapes of the as-built and heat-treated CoCrMo alloys with different, designed volume porosities of 0, 20, 40, 60, and 80%. A discernable variation in surface morphology towards the rougher surfaces and the change in color in the sample surfaces was found using heat treatment. The heat-treatment-induced varying color from slight yellow to dark green was ascribed to the presence of a chromium surface oxide, which was demonstrated by an obvious appearance of Cr_2O_3 [48–50], as shown in the XRD patterns in Figure 2.

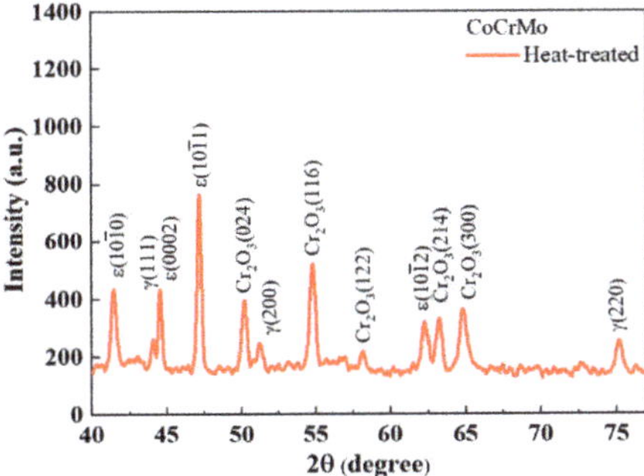

Figure 2. XRD profile on the surface of fully dense heat-treated CoCrMo alloy.

A martensitic transformation from an fcc to an hcp phase usually occurs in the CoCrMo alloys. Although the thermodynamically stable phase at room temperature is the hcp phase, the remaining fcc phase is mostly obtained in the CoCrMo alloys due to the sluggish transformation from the metastable fcc to the stable hcp under normal conditions [23,24]. The fcc to hcp transformation occurs more easily via rapid cooling, plastic deformation, and isothermal aging below the transformation temperature [23,24]. An examination of the degree of martensitic transformation via a heat treatment process was determined using XRD profiles in Figure 2. The as-built CoCrMo exhibited an obvious fcc phase with a lattice constant of 3.568 Å, while the heat-treated CoCrMo revealed the coexistence of residual fcc and hcp phases. The volume fractions of the hcp and fcc phases can be calculated as follows [25,51,52].

$$f_{hcp} = \frac{I(10\bar{1}1)_{hcp}}{I(10\bar{1}1)_{hcp} + 1.5I(200)_{fcc}} \tag{4}$$

$$f_{fcc} = 1 - f_{hcp} \tag{5}$$

where $I(10\bar{1}1)_{hcp}$ and $I(200)_{fcc}$ are the integrated intensities of the $(10\bar{1}1)_{hcp}$ and $(200)_{fcc}$ diffraction peaks for the hcp and fcc phases, respectively.

The calculated volume fraction of the fcc and hcp phases in the fully dense heat-treated CoCrMo alloy were 64 and 36%, respectively.

3.2. Relative Density of the Solid Structures in Both Conditions

Figure 3 shows the top-view surface in the as-built and heat-treated porous CoCrMo structures with different, designed volume porosities. The melt pool formed in different printing layers during SLM laser scanning was observed on the surface of fabricated samples, which was different from the smooth-designed model. The melt pool boundary became blurred, and the oxide layer, mainly composed of chromium oxide on the surface, made the surface rougher after heat treatment. In addition, there were also unmelted powder particles and spatter attached to the surface. When the melt pool was formed using a laser source, the excess heat energy would melt the nearby powders so they would adhere to the surface [39]. The distribution of pores was quite homogeneous in both conditions. Although the pore size on the surface of fabricated samples with a designed volume porosity of 20% was similar to that of the designed model, the pores were not as completely hollow in the vertical direction as those designed with the 3D model. The pores were obvious on the surface of fabricated samples with the designed volume porosity above 40%, and their pore sizes were evidently smaller than those of the designed model.

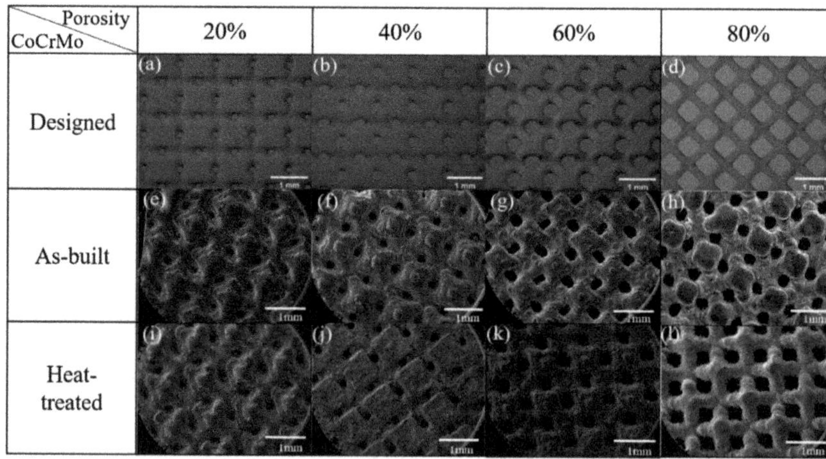

Porosity CoCrMo	20%	40%	60%	80%
Designed	(a)	(b)	(c)	(d)
As-built	(e)	(f)	(g)	(h)
Heat-treated	(i)	(j)	(k)	(l)

Figure 3. Top-view surface in (**a–d**) CAD-designed models, (**e–h**) as-built, and (**i–l**) heat-treated CoCrMo alloys with different, designed volume porosities.

Figure 4a describes the pore size in the designed, as-built, and heat-treated CoCrMo alloys. There was a negligible discrepancy in pore size between the as-built and heat-treated CoCrMo alloys. The difference in pore size between the fabricated and designed alloys increased with increasing designed volume porosity. As shown in the as-built CoCrMo alloys in Figure 4b, the pore fraction significantly increased with the increasing designed volume porosity.

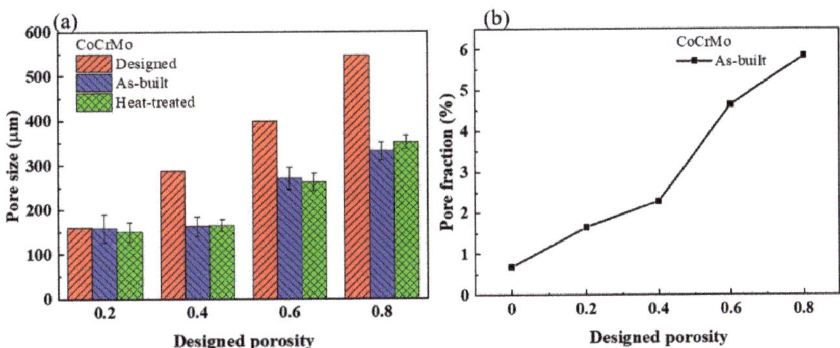

Figure 4. (**a**) Pore size in the designed, as-built, and heat-treated CoCrMo alloys with different, designed volume porosities. (**b**) Pore fraction in the as-built CoCrMo alloys with different, designed volume porosities.

Figure 5a describes the relative densities of the solids in the as-built and heat-treated CoCrMo alloys determined using Archimedes' method to examine the printing quality of the SLM process. The relative densities of the fully dense as-built and heat-treated samples were 99.5 and 99.6%, respectively. The relative densities of the heat-treated porous CoCrMo alloys altered from 97 to 98.4%, which were slightly lower than the variation from 98.3 to 99.2% of the as-built porous samples. Such high relative densities suggested a negligible existence of internal void defects on the bulk struts during the fabrication process. The relative density of the solid was reduced slightly with an increasing designed volume porosity within the range of 20–40%; however, the effect of the designed volume porosity on the relative density of the solid was generally trivial. The lower relative densities of the solid seen at the designed volume porosity of 20–40% implied that a small number of voids was confined in the solid strut.

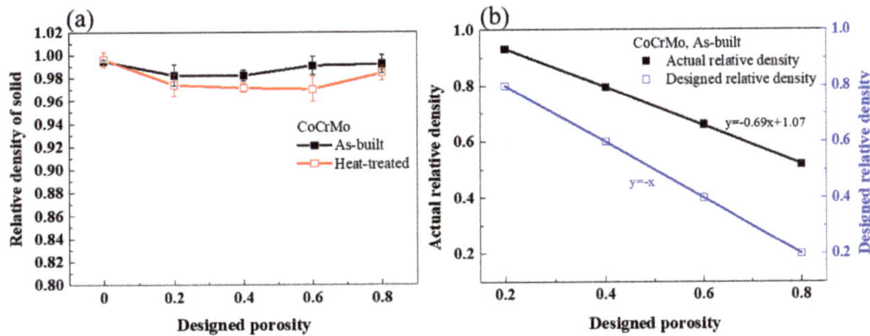

Figure 5. (**a**) The relative density of the solid in the as-built and heat-treated CoCrMo with various designed volume porosities. (**b**) The actual relative density and designed relative density in the as-built CoCrMo with various designed volume porosities.

The actual relative densities in the as-built porous CoCrMo were determined using Equation (3) and compared with the designed relative densities in Figure 5b. The actual relative densities in the as-built CoCrMo alloys were all higher than the designed relative densities, in accordance with previous work [53]. Moreover, the difference between the actual and designed relative densities became larger with increasing designed volume porosity. It was possibly related to a thinner strut and lower thermal conductivity, which may enlarge the melt pool and increase the width of the strut compared to the designed model. Table 1 lists the parameters of the as-built porous CoCrMo alloys.

Table 1. Parameters of the as-built porous CoCrMo alloys.

Designed Volume Porosity (%)	Designed Relative Density (%)	Actual Relative Density (%)	Relative Density of Solid (%)
20	80	93.3	98.3
40	60	79.5	98.2
60	40	66.2	99.1
80	20	52.1	99.2

3.3. Compressive Deformation in the As-Built and Heat-Treated Porous CoCrMo Structures

Figure 6 presents the macroscopic stress-strain curves of uniaxial compression tests in both conditions. The three distinct deformation stages of linear elasticity, plateau, and densification were seen at the designed volume porosity above 40% in both conditions, which was similar to the typical deformation of porous structures proposed by Gibson–Ashby [54]. Young's modulus was extracted in the initial stage of elastic deformation. The plastic deformation started to yield in the plateau region of the local collapse of pores, at which the strain significantly increased with a negligible variation in the stress. The plateau region was followed by a sharp increase in stress at the onset of the densification regime. The densification stages started at large strains of 0.65 and 0.74 in the heat-treated sample with a designed volume porosity of 60% and in the as-built CoCrMo with a designed volume porosity of 80%. In the present study, the SLM-built CoCrMo alloys with an actual porosity above 34% disclosed porous structures. Meanwhile, no obvious plateau and densification regions were obtained at a designed porosity below 40%, which is analogous to the deformation behavior of metallic solids. Such a drastic decline in stress after the elastic regime derived from the failure of struts in the porous structures owning high designed relative density.

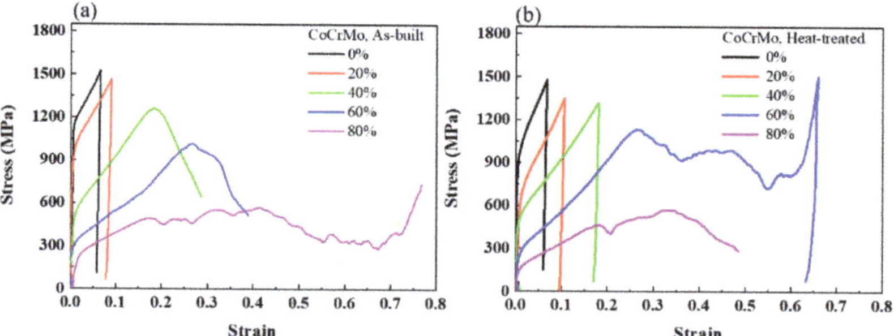

Figure 6. Compressive stress-strain curves of the (**a**) as-built and (**b**) heat-treated CoCrMo alloys with different, designed volume porosities.

Figure 7 shows the experimental and simulated values of Young's modulus and the compressive yield strength in both conditions with respect to the designed volume porosity. The yield strength drastically decreased from 1079 to 822 MPa, while there was a slight increase in Young's modulus from 191 to 198 GPa in the fully dense CoCrMo after heat treatment. The experimental values of Young's modulus and the yield strength were generally higher than the simulated values in both conditions and may be due to a higher actual relative density rather than the designed relative density. Increasing the designed volume porosity significantly decreased Young's modulus and the yield strength in both conditions. Furthermore, heat treatment resulted in a noticeable decrease in yield strengths but a slight increase in Young's moduli for all different, designed volume porosities of CoCrMo alloys. Such a remarkable discrepancy in yield strength between the as-built and heat-treated conditions became smaller with increasing designed volume porosity. Among the investigated CoCrMo alloys, the as-built and heat-treated CoCrMo porous structures with the designed volume porosity of 80% had Young's moduli of 17 and 29 GPa, respectively, while they possessed compressive yield strengths of 271 and 187 MPa, respectively. Compared to Young's modulus of 3–30 GPa and a yield strength of 193 MPa in the human cortical bone [18,55], the as-built and heat-treated CoCrMo porous structures with a designed volume porosity of 80% were the most appropriate implants for potential biomedical applications due to a very close match in their mechanical responses. In addition, heat treatment was more conducive to tailoring the mechanical performance of SLM-built porous structures, which was more similar to that of human cortical bone. The mechanical properties of CoCrMo structures could be effectively tuned using an adjustable, designed volume porosity fabricated via SLM.

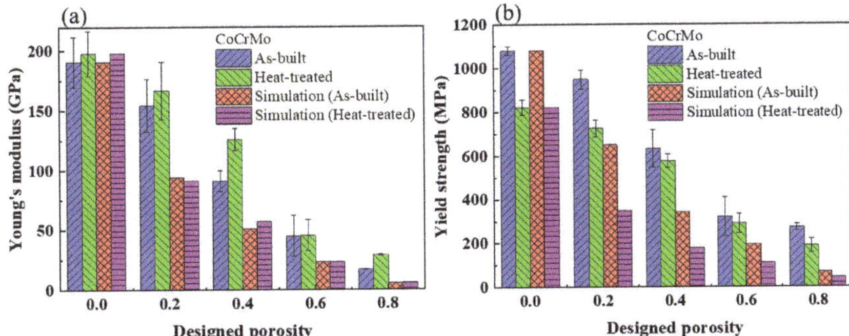

Figure 7. (**a**) Young's modulus and (**b**) yield strength in the as-built and heat-treated CoCrMo alloys determined from the experiment and simulation with different, designed volume porosities.

3.4. The Compressive Mechanical Properties Using the Gibson–Ashby Model and Finite Element Simulation

An analytical model proposed by Gibson–Ashby enables the effective prediction of the adjustable porosity for isotropic materials [54]. The compressive mechanical responses of porous CoCrMo structures were fitted using the Gibson–Ashby model to evaluate the degree of matching among the analytical, simulated, and experimental results in establishing a more suitable design of porous structures for potentially promising implants in biomedical applications. The compressive deformation of porous CoCrMo alloys was implemented using finite element simulation. The analytical prediction of an elastic modulus and yield strength was presented with the Gibson–Ashby model as follows [54].

$$\frac{E}{E_S} = C_1 \left(\frac{\rho}{\rho_S} \right)^n \tag{6}$$

$$\frac{\sigma}{\sigma_S} = C_5 \left(\frac{\rho}{\rho_S} \right)^m \tag{7}$$

where E and E_S are the elastic modulus of cellular and solid materials, respectively, σ and σ_S are the yield strength of cellular and solid materials, respectively, ρ and ρ_S are the density of cellular and solid materials, respectively, C_1 and C_5 are constants, and n and m are exponential factors. The general values of n and m are 2 and 1.5, respectively. C_1 and

C_5 are ideally 1; however, their variable values were experimentally determined from the best-fitting curve.

The relative elastic moduli E/E_S and relative yield strengths σ/σ_S obtained from the analytical, simulated, and experimental results versus a logarithmic scale of relative density were plotted in Figure 8. The fitted value of C_1 from the simulated and experimental results was in the range of 0.9–1.09. The variation in n is mainly related to the predominant deformation mode of strut-based cellular structures, which can be determined based on the lattice structure of repeating unit cells. Based on the Maxwell number of the bcc unit cell [56,57], the compressive deformation mechanism of the porous CoCrMo alloys was ascribed to the bending-dominated deformation proposed using the Gibson–Ashby model in which the exponent n is 2. The fitting curves of relative elastic modulus in both simulated and experimental results revealed linear relation with respect to the relative density, following the power law relationship. In Figure 8a, the exponent n extracted from the fitting curves of the simulated and experimental results was larger than that from the analytical Gibson–Ashby model. The fitting curve of the simulation analysis deviated from that of the Gibson–Ashby model, and it has an exponent of 2.58, which was close to the exponential value obtained in the simulation analysis of other bcc strut-based structures [45]. The fitting curve of as-built and heat-treated samples was closer to that of the simulation with an exponent of 3.56 and 3.05, respectively, which was far outside the expected range. The data of the simulation analysis and experiment were all well-fitted.

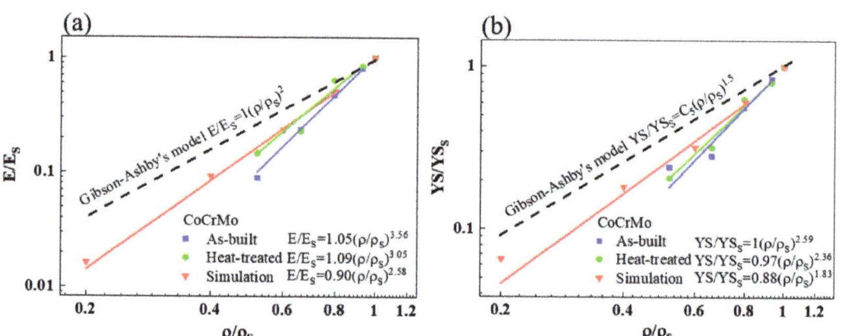

Figure 8. (**a**) The relative elastic modulus and (**b**) relative yield strength obtained from the analytical, simulated, and experimental results versus a logarithmic scale of relative density in the as-built and heat-treated CoCrMo alloys.

In Figure 8b, the relative yield strength versus relative density also followed a linear trend, which was similar to the relative elastic modulus versus relative density. The fitting curve of the relative yield strength in both simulated and experimental results also deviated from that proposed using the Gibson–Ashby model. The data of simulation analysis and heat-treated samples were better fitted, while there was a deviation between the fitting curve and data in the as-built samples. The exponent m obtained from the simulated, as-built, and heat-treated results were 1.83, 2.59, and 2.36, respectively, which was relatively greater than the 1.5 derived from the Gibson–Ashby model. The exponential value of 1.83 derived from simulation analysis was close to that of 1.97 in other bcc strut-based structures [45]. In general, a closer agreement was attained in the simulation analysis of bcc strut-based structures between this study and another previous study [45]. However, there was not a close correlation between the predicted and experimental values in the present study.

3.5. Microstructural Characterization in Both Conditions

The microstructures significantly govern the mechanical performance of SLM-built alloys. A considerably decreased yield strength in the heat-treated CoCrMo alloys was ascribed to the microstructural change after heat treatment. Figure 9 depicts OM images of the morphologies in the fully dense as-built and heat-treated CoCrMo alloys. In Figure 9a,c, two typical morphologies of melt pools in half-cylinder and stripe-like shapes were obvious in the plane parallel and perpendicular to the building direction, respectively, which was similarly seen in other SLM-built alloys. In Figure 9b,d, there was a disappearance of melt pools and a presence of grain boundaries with different grain sizes after heat treatment. In Figure 9b, most of the large grains were elongated with their long axes parallel to the building direction, implying an incomplete recrystallization process after heat treatment. The average length and width of elongated grains were 120 and 44 μm, respectively. Fine grains were also observed, which were newly formed grains in the initial stage of recrystallization. In Figure 9d, the grains were more likely to be equiaxed grains with an average size of 11 μm.

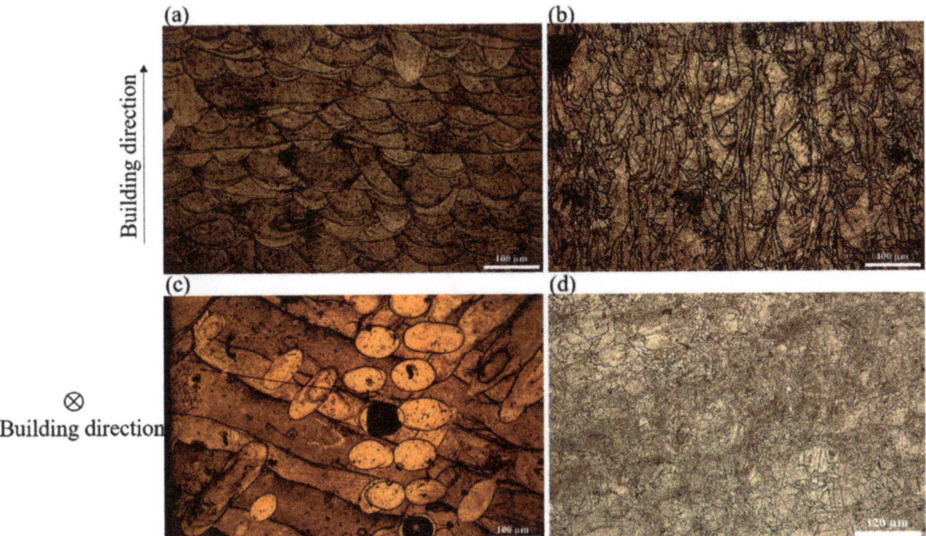

Figure 9. OM micrographs in the fully dense (**a**) as-built and (**b**) heat-treated CoCrMo alloys in the plane parallel to the building direction; (**c**) and (**d**) Those in the plane perpendicular to the building direction, respectively.

Figure 10 shows SEM images in the fully dense as-built and heat-treated CoCrMo alloys. A typical cell structure with an average size of 0.57 μm was obvious in the as-built CoCrMo, shown in Figure 10a. The energy-dispersive X-ray spectroscopy (EDS) analysis disclosed a different elemental distribution between the cell boundary and the cell. The cell boundary was found to be C-Mo rich, presumably ascribed to the $M_{23}C_6$ phase [29]. When the metal powder was heated and cooled rapidly, Mo with a high melting point was discharged to the cell boundary, and CoCr remained inside the cell [58]. In Figure 10b, the formation of precipitates was visible at both the grain boundaries and within the grains after heat treatment. The precipitates at the grain boundaries seemed to be slightly elongated along the grain boundaries, and their sizes were much greater than those inside the grains.

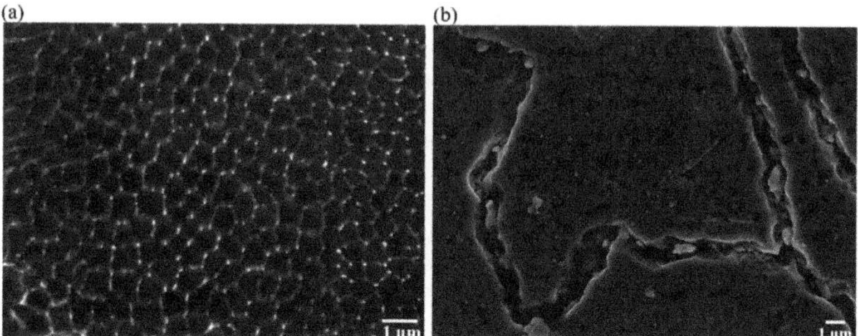

Figure 10. SEM images in the fully dense (**a**) as-built and (**b**) heat-treated CoCrMo alloys.

3.6. Strengthening Mechanisms

The mechanical properties of CoCrMo alloys were significantly altered in the as-built and heat-treated conditions, ascribed to their microstructural changes in governing the role of distinct strengthening behaviors, such as grain boundary strengthening, dislocation strengthening, and Orowan strengthening. The contribution of each strengthening mechanism to the calculated yield strength of CoCrMo alloys in both conditions was estimated.

The grain boundary strengthening is shown below [24,59].

$$\Delta\sigma_{GB} = kd^{-1/2} \tag{8}$$

where k is the Hall–Petch constant, and it has different values for the fcc and hcp phases. K is 400 [6] or 243.9 MPa $\mu m^{-1/2}$ [60] for the fcc or hcp phase, respectively. d is the average grain or cell size determined from SEM analysis.

The dislocation strengthening was described as follows [24].

$$\Delta\sigma_{dis} = \alpha MGb\rho_{dis}^{1/2} \tag{9}$$

where α is a dimensionless constant, M is the Taylor factor, G is the shear modulus, b is the Burgers vector, and ρ_{dis} is the dislocation density. α is 0.24 [61] or 0.1 [62] for the fcc or hcp phase, respectively. G is 78.4 [63] or 82.2 GPa [64] for the fcc or hcp phase, respectively. b is 0.1463 nm for both the fcc and hcp phases [14]. ρ_{dis} was obtained from CMWP fitting.

The Orowan mechanism was expressed as [65].

$$\Delta\sigma_{Orowan} = \frac{2Gb}{d_f}\left(\frac{6V_f}{\pi}\right)^{1/3} \tag{10}$$

where d_f is the average diameter of precipitates and V_f is the volume fraction of precipitates. d_f and V_f were determined from SEM analysis.

The calculated yield strength for the fcc or hcp phase was presented as

$$\sigma_i = M\tau_{CRSS} + \Delta\sigma_{GB} + \Delta\sigma_{dis} + \Delta\sigma_{Orowan} \tag{11}$$

where i is the fcc or hcp phase, M is the Taylor factor, and τ_{CRSS} is the critical resolved shear stress [66]. M is 2.57 [66] or 3.06 [25] for the fcc or hcp phase, respectively. τ_{CRSS} is 54 [67] or 184 MPa [62] for the fcc or hcp phase, respectively.

There was a negligible contribution of precipitation hardening, and only the fcc phase existed in the as-built CoCrMo alloys. Meanwhile, since there existed two phases of fcc and hcp after heat treatment, the calculated yield strength of heat-treated CoCrMo alloys could be estimated with the rule of mixture as shown below [24].

$$\sigma_y = f_{hcp}\sigma_{hcp} + \left(1 - f_{hcp}\right)\sigma_{fcc} \tag{12}$$

where f_{hcp} is the volume fraction of hcp grains, and σ_{hcp} and σ_{fcc} are the strengths of hcp and fcc grains, respectively.

Figure 11 describes the contributing strength values and calculated yield strengths compared with the measured yield strengths in the fully dense as-built and heat-treated CoCrMo alloys. The calculated yield strengths in the as-built and heat-treated CoCrMo alloys were 1055 and 790 MPa, respectively, which were in very good accordance with the measured yield strengths of 1079 and 822 MPa, respectively. Both the strength values of grain boundary strengthening and dislocation strengthening were reduced after heat treatment. The disappearance of cellular structure, the increase in grain size, and the decrease in dislocation density were mainly responsible for the decreased yield strength after heat treatment.

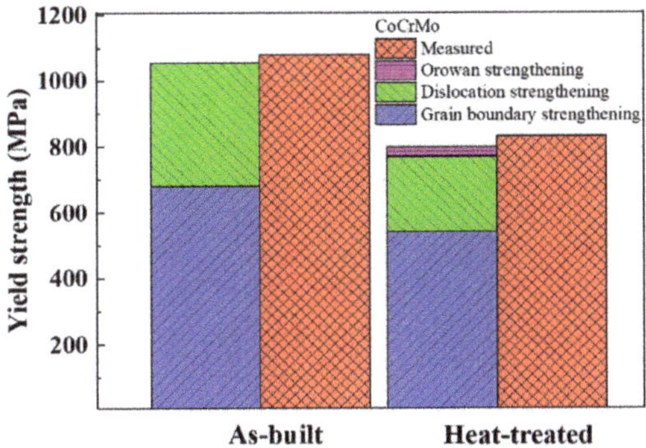

Figure 11. The measured and calculated yield strengths consisting of the grain boundary strengthening, dislocation strengthening, and Orowan strengthening in the fully dense as-built and heat-treated CoCrMo alloys.

4. Discussion

The Gibson–Ashby model is one of the most popular theoretical predictions to evaluate the correlation between porous structures and their mechanical performance. Clarifying reasonable factors governing the difference between the predicted and experimental results is necessary to establish a better design for SLM-built porous CoCrMo structures. The predicted values from simulation and analytical analysis underestimated the elastic modulus and yield strength of the CoCrMo alloys as compared to the experimental values. Higher exponential values derived from the experimental data of elastic modulus and yield strength were presumably attributed to the structural characteristics of the bcc strut-based structures. The strut was not parallel to the compression direction, which was different from the Gibson–Ashby model. Thus, the resistance to load deformation is weak and results in larger exponential factors of elastic modulus and yield strength [68]. Another possible reason is the presence of defects during the SLM process, causing the discrepancy between the designed and actual relative density [69]. Due to the increased surface area of the strut, the printing process causes a partial melting of the loose powder beneath it and bonds it to the strut surfaces, resulting in the increased weight and higher actual relative density of the specimen but a negligible contribution to the mechanical strength [45,69]. In addition, the corrugation and increased surface roughness of the strut caused by partially melted powder possibly induce stress concentration and thus lead to a lower elastic modulus as well as a lower yield strength of the SLM-built alloys [69]. As the relative density decreases, the thinner the strut, the slower the cooling rate, and the coarser the microstructure, which acts as another reasonable factor in reducing the yield strength besides the density effect [70]. Such a lower yield strength results in an increase in the exponential factor of the experimental data. Although the elastic modulus and yield strength were not well fitted using the Gibson–Ashby model, the predicted values could still be referred to for future designs of SLM-built porous CoCrMo alloys with adjustable mechanical properties.

5. Conclusions

The role of different, designed volume porosities and heat treatment processes on the mechanical properties of SLM-built CoCrMo alloys were investigated. The SLM-built CoCrMo with an actual porosity above 34% exhibited porous structures. An optimal actual porosity of 48% resulted in appropriate mechanical responses of the SLM-built CoCrMo structures compared to those of the human bone. Furthermore, the heat treatment process was found to be more beneficial in tailoring Young's modulus and the yield strength of the SLM-built CoCrMo alloys with a minimal stress shielding effect. Possible explanations for the underestimated exponential factors from simulation and analytical analysis compared to the experimental values were reported. Our findings suggest the optimal design of bcc lattice-structure-based CoCrMo alloys for a closer match in mechanical properties between the porous SLM-built CoCrMo implants and bone tissue for potential biomedical applications.

Author Contributions: Conceptualization, E.-W.H.; formal analysis, T.-N.L. and K.-M.C.; investigation, K.-M.C. and C.-H.T.; resources, P.-I.T., M.-H.W. and C.-C.H.; writing—original draft preparation, T.-N.L.; writing—review and editing, J.J. and E.-W.H.; supervision, E.-W.H.; project administration, E.-W.H. All authors have read and agreed to the published version of the manuscript.

Funding: The authors are grateful to the support of the National Science and Technology Council (NSTC), Taiwan, under Grant No. 110-2224-E-007-001, 108-2221-E-009-131-MY4, and 111-2811-E-A49-503. This work was financially supported by the "Center for the Semiconductor Technology Research" from The Featured Areas Research Center Program within the framework of the Higher Education Sprout Project by the Ministry of Education (MOE) in Taiwan. Also supported in part by the National Science and Technology Council, Taiwan, under Grant No. NSTC 111-2634-F-A49-008-. This work was supported by the Higher Education Sprout Project of the National Yang Ming Chiao Tung University and Ministry of Education (MOE), Taiwan.

Informed Consent Statement: Not applicable.

Data Availability Statement: The data presented in this study are available on request from the corresponding author.

Acknowledgments: The authors would like to thank ITRI for providing the SLM-built CoCrMo alloys.

Conflicts of Interest: The authors declare no conflict of interest.

Nomenclature

bcc	Body-centered cubic
SLM	Selective laser melting
hcp	Hexagonal closest packed
Ta	Tantalum
Ti	Titanium
Co	Cobalt
CoCrMo	Cobalt-chromium-molybdenum
3D	Three-dimensional
CADITRI	Computer aided design Industrial Technology Research Institute
OM	Optical microscope
SEM	Scanning electron microscopy
ρ_f	Density of foam
ρ_S	Density of solid structure
M_{porous}	Weight of the porous sample
V_{porous}	Volume of the porous sample

ρ_S	Density of solid structure
ρ_w	Density of water
w_a	Weight of the sample in air
w_w	Weight of the sample in water
ρ^*	Relative density of material
$\rho_{measured}$	Measured density
$\rho_{theoretical}$	Theoretical bulk density
f_{hcp}	Volume fraction of hcp
f_{fcc}	Volume fraction of fcc
$I(10\bar{1}1)_{hcp}$	Integrated intensity of the $(10\bar{1}1)_{hcp}$ peaks for the hcp
$I(200)_{fcc}$	Integrated intensity of the $(200)_{fcc}$ peaks for the fcc
E/E_S	Relative elastic modulus
σ/σ_S	Relative yield strength
E	Elastic modulus of cellular material
E_S	Elastic modulus of solid material
σ	Yield strength of cellular material
σ_S	Yield strength of solid material
ρ	Density of cellular material
ρ_S	Density of solid material
C_1	Constant
C_5	Constant
m	Exponential factor
n	Exponential factor
EDS	Energy-dispersive X-ray spectroscopy
$\Delta\sigma_{GB}$	Grain boundary strengthening
k	Hall-Petch constant
d	Average grain or cell size
$\Delta\sigma_{dis}$	Dislocation strengthening
α	Dimensionless constant
M	Taylor factor
G	Shear modulus
b	Burgers vector
ρ_{dis}	Dislocation density
$\Delta\sigma_{Orowan}$	Orowan strengthening
d_f	Average diameter of precipitates
V_f	Volume fraction of precipitates
i	fcc or hcp phase
τ_{CRSS}	Critical resolved shear stress
σ_{hcp}	Strength of hcp
σ_{fcc}	Strength of fcc

References

1. Lim, Y.Y.; Zaidi, A.M.A.; Miskon, A. Composing On-Program Triggers and On-Demand Stimuli into Biosensor Drug Carriers in Drug Delivery Systems for Programmable Arthritis Therapy. *Pharmaceuticals* **2022**, *15*, 1330. [CrossRef]
2. Kenel, C.; Casati, N.P.M.; Dunand, D.C. 3D ink-extrusion additive manufacturing of CoCrFeNi high-entropy alloy micro-lattices. *Nat. Commun.* **2019**, *10*, 904. [CrossRef]
3. Black, J. Corrosion and Degradation. In *Orthopedic Biomaterials in Research and Practice*; Churchill Livingstone: New York, NY, USA, 1988; pp. 235–266.
4. Lin, H.-Y.; Bumgardner, J.D. Changes in the surface oxide composition of Co–Cr–Mo implant alloy by macrophage cells and their released reactive chemical species. *Biomaterials* **2004**, *25*, 1233–1238. [CrossRef]
5. Chiba, A.; Kumagai, K.; Nomura, N.; Miyakawa, S. Pin-on-disk wear behavior in a like-on-like configuration in a biological environment of high carbon cast and low carbon forged Co–29Cr–6Mo alloys. *Acta Mater.* **2007**, *55*, 1309–1318. [CrossRef]
6. Yamanaka, K.; Mori, M.; Chiba, A. Enhanced Mechanical Properties of As-Forged Co-Cr-Mo-N Alloys with Ultrafine-Grained Structures. *Met. Mater. Trans. A* **2012**, *43*, 5243–5257. [CrossRef]
7. Mischler, S.; Muñoz, A.I. Wear of CoCrMo alloys used in metal-on-metal hip joints: A tribocorrosion appraisal. *Wear* **2013**, *297*, 1081–1094. [CrossRef]
8. Zhang, X.; Li, Y.; Tang, N.; Onodera, E.; Chiba, A. Corrosion behaviour of CoCrMo alloys in 2 wt% sulphuric acid solution. *Electrochim. Acta* **2014**, *125*, 543–555. [CrossRef]

9. Henriques, B.; Bagheri, A.; Gasik, M.; Souza, J.; Carvalho, O.; Silva, F.; Nascimento, R.M.D. Mechanical properties of hot pressed CoCrMo alloy compacts for biomedical applications. *Mater. Des.* **2015**, *83*, 829–834. [CrossRef]

10. Hussein, M.A.; Mohammed, A.S.; Al-Aqeeli, N. Wear Characteristics of Metallic Biomaterials: A Review. *Materials* **2015**, *8*, 2749–2768. [CrossRef]

11. Muthaiah, V.S.; Indrakumar, S.; Suwas, S.; Chatterjee, K. Surface engineering of additively manufactured titanium alloys for enhanced clinical performance of biomedical implants: A review of recent developments. *Bioprinting* **2022**, *25*, e00180. [CrossRef]

12. Giacchi, J.; Morando, C.; Fornaro, O.; Palacio, H. Microstructural characterization of as-cast biocompatible Co–Cr–Mo alloys. *Mater. Charact.* **2011**, *62*, 53–61. [CrossRef]

13. Bordin, A.; Ghiotti, A.; Bruschi, S.; Facchini, L.; Bucciotti, F. Machinability Characteristics of Wrought and EBM CoCrMo Alloys. *Procedia CIRP* **2014**, *14*, 89–94. [CrossRef]

14. Yamanaka, K.; Mori, M.; Sato, S.; Chiba, A. Stacking-fault strengthening of biomedical Co-Cr-Mo alloy via multipass thermomechanical processing. *Sci. Rep.* **2017**, *7*, 10808. [CrossRef] [PubMed]

15. Agustini, P.; Iwan, S. Microstructural and Mechanical Characterization of As-Cast Co-Cr-Mo Alloys with Various Content of Carbon and Nitrogen. *Mater. Sci. Forum* **2020**, *988*, 206–211. [CrossRef]

16. Yan, C.; Hao, L.; Hussein, A.; Raymont, D. Evaluations of cellular lattice structures manufactured using selective laser melting. *Int. J. Mach. Tools Manuf.* **2012**, *62*, 32–38. [CrossRef]

17. Parthasarathy, J.; Starly, B.; Raman, S. A design for the additive manufacture of functionally graded porous structures with tailored mechanical properties for biomedical applications. *J. Manuf. Process.* **2011**, *13*, 160–170. [CrossRef]

18. Hazlehurst, K.; Wang, C.J.; Stanford, M. Evaluation of the stiffness characteristics of square pore CoCrMo cellular structures manufactured using laser melting technology for potential orthopaedic applications. *Mater. Des.* **2013**, *51*, 949–955. [CrossRef]

19. Emmelmann, C.; Scheinemann, P.; Munsch, M.; Seyda, V. Laser Additive Manufacturing of Modified Implant Surfaces with Osseointegrative Characteristics. *Phys. Procedia* **2011**, *12*, 375–384. [CrossRef]

20. Tsai, P.-I.; Lam, T.-N.; Wu, M.-H.; Tseng, K.-Y.; Chang, Y.-W.; Sun, J.-S.; Li, Y.-Y.; Lee, M.-H.; Chen, S.-Y.; Chang, C.-K.; et al. Multi-scale mapping for collagen-regulated mineralization in bone remodeling of additive manufacturing porous implants. *Mater. Chem. Phys.* **2019**, *230*, 83–92. [CrossRef]

21. Lim, Y.Y.; Miskon, A.; Zaidi, A.M.A. Structural Strength Analyses for Low Brass Filler Biomaterial with Anti-Trauma Effects in Articular Cartilage Scaffold Design. *Materials* **2022**, *15*, 4446. [CrossRef]

22. Balagna, C.; Spriano, S.; Spriano, S. Characterization of Co–Cr–Mo alloys after a thermal treatment for high wear resistance. *Mater. Sci. Eng. C* **2012**, *32*, 1868–1877. [CrossRef] [PubMed]

23. López, H.; Saldivar-Garcia, A. Martensitic Transformation in a Cast Co-Cr-Mo-C Alloy. *Met. Mater. Trans. A* **2007**, *39*, 8–18. [CrossRef]

24. Roudnická, M.; Kubásek, J.; Pantělejev, L.; Molnárová, O.; Bigas, J.; Drahokoupil, J.; Paloušek, D.; Vojtěch, D. Heat treatment of laser powder-bed-fused Co–28Cr–6Mo alloy to remove its microstructural instability by massive FCC→HCP transformation. *Addit. Manuf.* **2021**, *47*, 102265. [CrossRef]

25. Wang, Z.; Tang, S.; Scudino, S.; Ivanov, Y.; Qu, R.; Wang, D.; Yang, C.; Zhang, W.; Greer, A.; Eckert, J.; et al. Additive manufacturing of a martensitic Co–Cr–Mo alloy: Towards circumventing the strength–ductility trade-off. *Addit. Manuf.* **2021**, *37*, 101725. [CrossRef]

26. Song, C.; Park, H.; Seong, H.; Pez, H.F.L. Development of athermal and isothermal ε-martensite in atomized Co-Cr-Mo-C implant alloy powders. *Metall. Mater. Trans. A* **2006**, *37*, 3197. [CrossRef]

27. Koizumi, Y.; Suzuki, S.; Yamanaka, K.; Lee, B.-S.; Sato, K.; Li, Y.; Kurosu, S.; Matsumoto, H.; Chiba, A. Strain-induced martensitic transformation near twin boundaries in a biomedical Co–Cr–Mo alloy with negative stacking fault energy. *Acta Mater.* **2013**, *61*, 1648–1661. [CrossRef]

28. Mori, M.; Yamanaka, K.; Chiba, A. Effect of cold rolling on phase decomposition in biomedical Co–29Cr–6Mo–0.2N alloy during isothermal heat treatment at 1073 K. *J. Alloys Compd.* **2014**, *612*, 273–279. [CrossRef]

29. Roudnicka, M.; Bigas, J.; Molnarova, O.; Palousek, D.; Vojtech, D. Different Response of Cast and 3D-Printed Co-Cr-Mo Alloy to Heat Treatment: A Thorough Microstructure Characterization. *Metals* **2021**, *11*, 687. [CrossRef]

30. Bawane, K.K.; Srinivasan, D.; Banerjee, D. Microstructural Evolution and Mechanical Properties of Direct Metal Laser-Sintered (DMLS) CoCrMo After Heat Treatment. *Met. Mater. Trans. A* **2018**, *49*, 3793–3811. [CrossRef]

31. Sames, W.J.; List, F.A.; Pannala, S.; Dehoff, R.R.; Babu, S.S. The metallurgy and processing science of metal additive manufacturing. *Int. Mater. Rev.* **2016**, *61*, 315–360. [CrossRef]

32. Gao, W.; Zhang, Y.; Ramanujan, D.; Ramani, K.; Chen, Y.; Williams, C.B.; Wang, C.C.L.; Shin, Y.C.; Zhang, S.; Zavattieri, P.D. The status, challenges, and future of additive manufacturing in engineering. *Comput. Aided Des.* **2015**, *69*, 65–89. [CrossRef]

33. Murr, L.E.; Gaytan, S.M.; Medina, F.; Lopez, H.; Martinez, E.; Machado, B.I.; Hernandez, D.H.; Lopez, M.I.; Wicker, R.B.; Bracke, J. Next-generation biomedical implants using additive manufacturing of complex, cellular and functional mesh arrays. *Philos. Trans. R. Soc. A Math. Phys. Eng. Sci.* **2010**, *368*, 1999–2032. [CrossRef]

34. Tseng, J.-C.; Huang, W.-C.; Chang, W.; Jeromin, A.; Keller, T.F.; Shen, J.; Chuang, A.C.; Wang, C.-C.; Lin, B.-H.; Amalia, L.; et al. Deformations of Ti-6Al-4V additive-manufacturing-induced isotropic and anisotropic columnar structures: Insitu measurements and underlying mechanisms. *Addit. Manuf.* **2020**, *35*, 101322. [CrossRef] [PubMed]

35. Lam, T.-N.; Trinh, M.-G.; Huang, C.-C.; Kung, P.-C.; Huang, W.-C.; Chang, W.; Amalia, L.; Chin, H.-H.; Tsou, N.-T.; Shih, S.-J.; et al. Investigation of Bone Growth in Additive-Manufactured Pedicle Screw Implant by Using Ti-6Al-4V and Bioactive Glass Powder Composite. *Int. J. Mol. Sci.* **2020**, *21*, 7438. [CrossRef] [PubMed]

36. Huang, E.-W.; Lee, W.-J.; Singh, S.S.; Kumar, P.; Lee, C.-Y.; Lam, T.-N.; Chin, H.-H.; Lin, B.-H.; Liaw, P.K. Machine-learning and high-throughput studies for high-entropy materials. *Mater. Sci. Eng. R Rep.* **2022**, *147*, 100645. [CrossRef]

37. Herzog, D.; Seyda, V.; Wycisk, E.; Emmelmann, C. Additive manufacturing of metals. *Acta Mater.* **2016**, *117*, 371–392. [CrossRef]

38. Tammas-Williams, S.; Todd, I. Design for additive manufacturing with site-specific properties in metals and alloys. *Scr. Mater.* **2017**, *135*, 105–110. [CrossRef]

39. Maconachie, T.; Leary, M.; Lozanovski, B.; Zhang, X.; Qian, M.; Faruque, O.; Brandt, M. SLM lattice structures: Properties, performance, applications and challenges. *Mater. Des.* **2019**, *183*, 108137. [CrossRef]

40. Alabort, E.; Barba, D.; Reed, R.C. Design of metallic bone by additive manufacturing. *Scr. Mater.* **2019**, *164*, 110–114. [CrossRef]

41. Azidin, A.; Taib, Z.A.M.; Harun, W.S.W.; Ghani, S.A.C.; Faisae, M.F.; Omar, M.A.; Ramli, H. Investigation of mechanical properties for open cellular structure CoCrMo alloy fabricated by selective laser melting process. *IOP Conf. Ser. Mater. Sci. Eng.* **2015**, *100*, 012033. [CrossRef]

42. Ghani, S.A.C.; Mohamed, S.R.; Sha'Ban, M.; Harun, W.S.W.; Noar, N.A.Z.M. Experimental investigation of biological and mechanical properties of CoCrMo based selective laser melted metamaterials for bone implant manufacturing. *Procedia CIRP* **2020**, *89*, 79–91. [CrossRef]

43. Lim, Y.Y.; Miskon, A.; Zaidi, A.M.A.; Ahmad, M.M.H.M.; Abu Bakar, M. Structural Characterization Analyses of Low Brass Filler Biomaterial for Hard Tissue Implanted Scaffold Applications. *Materials* **2022**, *15*, 1421. [CrossRef] [PubMed]

44. Song, C.; Zhang, M.; Yang, Y.; Wang, D.; Jia-Kuo, Y. Morphology and properties of CoCrMo parts fabricated by selective laser melting. *Mater. Sci. Eng. A* **2018**, *713*, 206–213. [CrossRef]

45. Al-Ketan, O.; Abu Al-Rub, R.K.; Rowshan, R. The effect of architecture on the mechanical properties of cellular structures based on the IWP minimal surface. *J. Mater. Res.* **2018**, *33*, 343–359. [CrossRef]

46. Ghani, S.A.C.; Harun, W.S.W.; Taib, Z.A.M.; Ab Rashid, F.F.; Hazlen, R.M.; Omar, M.A. Finite Element Analysis of Porous Medical Grade Cobalt Chromium Alloy Structures Produced by Selective Laser Melting. *Adv. Mater. Res.* **2016**, *1133*, 113–118. [CrossRef]

47. Martinez-Marquez, D.; Delmar, Y.; Sun, S.; Stewart, R.A. Exploring Macroporosity of Additively Manufactured Titanium Metamaterials for Bone Regeneration with Quality by Design: A Systematic Literature Review. *Materials* **2020**, *13*, 4794. [CrossRef]

48. Abdullah, M.M.; Rajab, F.M.; Al-Abbas, S.M. Structural and optical characterization of Cr_2O_3 nanostructures: Evaluation of its dielectric properties. *AIP Adv.* **2014**, *4*, 027121. [CrossRef]

49. Yeom, H.; Maier, B.; Johnson, G.; Dabney, T.; Lenling, M.; Sridharan, K. High temperature oxidation and microstructural evolution of cold spray chromium coatings on Zircaloy-4 in steam environments. *J. Nucl. Mater.* **2019**, *526*, 151737. [CrossRef]

50. Huynh, V.; Ngo, N.K.; Golden, T.D. Surface Activation and Pretreatments for Biocompatible Metals and Alloys Used in Biomedical Applications. *Int. J. Biomater.* **2019**, *2019*, 3806504. [CrossRef]

51. Sage, M.; Guillaud, C. Méthode d'analyse quantitative des variétés allotropiques du cobalt par les rayons X. *Revue de Métallurgie* **2017**, *47*, 139–145. [CrossRef]

52. Ramirez-Ledesma, A.; Lopez-Molina, E.; Lopez, H.; Juarez-Islas, J. Athermal ε-martensite transformation in a Co–20Cr alloy: Effect of rapid solidification on plate nucleation. *Acta Mater.* **2016**, *111*, 138–147. [CrossRef]

53. Van Hooreweder, B.; Lietaert, K.; Neirinck, B.; Lippiatt, N.; Wevers, M. CoCr F75 scaffolds produced by additive manufacturing: Influence of chemical etching on powder removal and mechanical performance. *J. Mech. Behav. Biomed. Mater.* **2017**, *68*, 216–223. [CrossRef] [PubMed]

54. Gibson, L.J.; Ashby, M.F. *Cellular Solids: Structure and Properties*; Cambridge University Press: Cambridge, UK, 1999.

55. Wang, X.; Xu, S.; Zhou, S.; Xu, W.; Leary, M.; Choong, P.; Qian, M.; Brandt, M.; Xie, Y.M. Topological design and additive manufacturing of porous metals for bone scaffolds and orthopaedic implants: A review. *Biomaterials* **2016**, *83*, 127–141. [CrossRef] [PubMed]

56. Deshpande, V.S.; Fleck, N.A.; Ashby, M.F. Effective properties of the octet-truss lattice material. *J. Mech. Phys. Solids* **2001**, *49*, 1747–1769. [CrossRef]

57. Leary, M.; Mazur, M.; Williams, H.; Yang, E.; Alghamdi, A.; Lozanovski, B.; Zhang, X.; Shidid, D.; Farahbod-Sternahl, L.; Witt, G.; et al. Inconel 625 lattice structures manufactured by selective laser melting (SLM): Mechanical properties, deformation and failure modes. *Mater. Des.* **2018**, *157*, 179–199. [CrossRef]

58. Prashanth, K.G.; Eckert, J. Formation of metastable cellular microstructures in selective laser melted alloys. *J. Alloys Compd.* **2017**, *707*, 27–34. [CrossRef]

59. Lam, T.-N.; Lee, A.; Chiu, Y.-R.; Kuo, H.-F.; Kawasaki, T.; Harjo, S.; Jain, J.; Lee, S.Y.; Huang, E.-W. Estimating fine melt pool, coarse melt pool, and heat affected zone effects on the strengths of additive manufactured AlSi10Mg alloys. *Mater. Sci. Eng. A* **2022**, *856*, 143961. [CrossRef]

60. Fleurier, G.; Hug, E.; Martinez, M.; Dubos, P.-A.; Keller, C. Size effects and Hall–Petch relation in polycrystalline cobalt. *Philos. Mag. Lett.* **2015**, *95*, 122–130. [CrossRef]

61. Hansen, N.; Huang, X. Microstructure and flow stress of polycrystals and single crystals. *Acta Mater.* **1998**, *46*, 1827–1836. [CrossRef]

62. Matsumoto, H.; Koizumi, Y.; Ohashi, T.; Lee, B.-S.; Li, Y.; Chiba, A. Microscopic mechanism of plastic deformation in a polycrystalline Co–Cr–Mo alloy with a single hcp phase. *Acta Mater.* **2014**, *64*, 1–11. [CrossRef]

63. Mori, M.; Yamanaka, K.; Sato, S.; Wagatsuma, K.; Chiba, A. Microstructures and Mechanical Properties of Biomedical Co-29Cr-6Mo-0.14N Alloys Processed by Hot Rolling. *Met. Mater. Trans. A* **2012**, *43*, 3108–3119. [CrossRef]

64. Sargent, P.; Malakondaiah, G.; Ashby, M. A deformation map for cobalt. *Scr. Met.* **1983**, *17*, 625–629. [CrossRef]

65. Liu, G.; Zhang, G.J.; Jiang, F.; Ding, X.D.; Sun, Y.J.; Sun, J.; Ma, E. Nanostructured high-strength molybdenum alloys with unprecedented tensile ductility. *Nat. Mater.* **2013**, *12*, 344–350. [CrossRef] [PubMed]

66. Kim, K.-S.; Hwang, J.-W.; Lee, K.-A. Effect of building direction on the mechanical anisotropy of biocompatible Co–Cr–Mo alloy manufactured by selective laser melting process. *J. Alloys Compd.* **2020**, *834*, 155055. [CrossRef]

67. Hagihara, K.; Nakano, T.; Sasaki, K. Anomalous strengthening behavior of Co–Cr–Mo alloy single crystals for biomedical applications. *Scr. Mater.* **2016**, *123*, 149–153. [CrossRef]

68. McKown, S.; Shen, Y.; Brookes, W.; Sutcliffe, C.; Cantwell, W.; Langdon, G.; Nurick, G.; Theobald, M. The quasi-static and blast loading response of lattice structures. *Int. J. Impact Eng.* **2008**, *35*, 795–810. [CrossRef]

69. Van Bael, S.; Kerckhofs, G.; Moesen, M.; Pyka, G.; Schrooten, J.; Kruth, J. Micro-CT-based improvement of geometrical and mechanical controllability of selective laser melted Ti6Al4V porous structures. *Mater. Sci. Eng. A* **2011**, *528*, 7423–7431. [CrossRef]

70. Cheng, X.; Li, S.; Murr, L.; Zhang, Z.; Hao, Y.; Yang, R.; Medina, F.; Wicker, R. Compression deformation behavior of Ti–6Al–4V alloy with cellular structures fabricated by electron beam melting. *J. Mech. Behav. Biomed. Mater.* **2012**, *16*, 153–162. [CrossRef]

materials

Editorial

Recent Progress in Additive Manufacturing of Alloys and Composites

Feiyang Gao [1], Haizhou Lu [2], Chao Zhao [3], Haokai Dong [1] and Lehua Liu [1,*]

1   National Engineering Research Center of Near-Net-Shape Forming for Metallic Materials, Guangdong Provincial Key Laboratory for Processing and Forming of Advanced Metallic Materials, South China University of Technology, Guangzhou 510640, China; gfy20018@163.com (F.G.); donghk@scut.edu.cn (H.D.)
2   School of Mechatronic Engineering, Guangdong Polytechnic Normal University, Guangzhou 510665, China; hzlu@gpnu.edu.cn
3   School of Materials Science and Engineering, Huazhong University of Science and Technology, Wuhan 430074, China; czhao@hust.edu.cn
*   Correspondence: liulh@scut.edu.cn

1. Introduction

Additive manufacturing, commonly referred to as 3D printing, is a fabrication method characterized by a layer-by-layer deposition process [1]. Its procedural framework encompasses several sequential stages, including design conceptualization, computer-aided modeling, slicing of digital designs into printable layers, physical printing, and subsequent post-processing treatments. In contrast to conventional manufacturing methodologies, additive manufacturing presents notable advantages such as enhanced geometric versatility, reduced material wastage, and expedited production cycles [2]. Furthermore, the distinctive attributes of additively manufactured alloys and composites, which include a profusion of metastable microstructures and exceptional material properties, have attracted considerable scholarly and industrial attention. These attributes are primarily ascribed to the inherent characteristics of additive manufacturing processes, including rapid cooling rates, large thermal gradients, and intricate thermal cycling histories. As a result, additive manufacturing has emerged as a focal point of interdisciplinary research endeavors, garnering widespread interest and engagement from academic and industrial stakeholders alike.

Currently, additive manufacturing is extensively applied in the fabrication of steels [3], nonferrous alloys [4], and metal matrix composites [5], meeting the stringent performance requirements across diverse sectors including aerospace, automotive, electronics, medical, military, and architecture. However, the additive manufacturing of alloys and composite materials still faces significant challenges. These challenges encompass an inadequate understanding of the complex interactions between processing parameters, the resulting microstructures, and the ensuing material properties. Additionally, elevated production costs, the prevalence of defects, and a lack of established theories elucidating the principles of physical metallurgy governing additive manufacturing processes further complicate the advancement and optimization of these technologies.

To further advance additive manufacturing technology, Dr. Liu et al. recently curated a Special Issue entitled "Additive Manufacturing of Alloys and Composites". This Special Issue focuses on advanced materials and related processes, aiming to explore the mechanisms governing microstructural evolution and property enhancement in additive manufacturing. Ultimately, ten contributions showcasing significant advancements in this domain were selected for inclusion. These articles cover additively manufactured stainless steels, superalloys, CoCrMo alloys, and metal matrix composites. Through these publications, this Special Issue not only presents the latest developments in high-performance

Citation: Gao, F.; Lu, H.; Zhao, C.; Dong, H.; Liu, L. Recent Progress in Additive Manufacturing of Alloys and Composites. *Materials* **2024**, *17*, 2905. https://doi.org/10.3390/ma17122905

Received: 20 May 2024
Accepted: 12 June 2024
Published: 13 June 2024

additive manufacturing materials but also paves the way for future advancements in additive manufacturing technology and the development of advanced materials.

2. Contributions

2.1. Alloys

Stainless steel, with its excellent mechanical properties and corrosion resistance, is gaining increasing attention in both scientific and technological fields. Chang et al. [6] successfully prepared complex 15-5PH stainless steel components by combining powder metallurgy with fused deposition modeling (FDM). Their results demonstrated optimal material fluidity at 285 °C during the FDM process and a solvent debinding rate of 98.7% achieved at 75 °C over 24 h. Furthermore, during the sintering process, the relative density of the sintered parts reached 95.83% at a sintering temperature of 1390 °C. This study provides valuable guidance for optimizing the FDM, debinding, and sintering processes.

In contemporary industrial applications, duplex stainless steels (DSSs) are extensively utilized across various sectors. However, the challenges in producing additive-manufactured DSSs are compounded by high levels of nitrogen (N), chromium (Cr), molybdenum (Mo), and other alloying elements. To address these issues, He et al. [7] developed a novel low-N 25Cr-type DSS. This novel low-N 25Cr-type DSS, fabricated using the Laser Powder Bed Fusion (L-PBF) method, exhibited a yield strength of 712 MPa and an elongation of 27.5%. Moreover, this study revealed that solution treatment at 1200 °C leads to the formation of discrete and refined austenite precipitates at ferrite grain boundaries, thereby enhancing strength and ductility. In their subsequent work [8], they transferred their eyes to the corrosion resistance investigation of low-N 25Cr-type duplex stainless steel prepared by the L-PBF method and solution treatment and analyzed the mechanism behind corrosion resistance enhancement. The results showed after solution treatment at 1200 °C for 1 h, the residual thermal stress in the specimen was eliminated and the Cr content in the ferrite phase increased, leading to an improvement in corrosion resistance.

To mitigate the stress shielding phenomenon in bone implants, Lam et al. [9] employed a selective laser melting (SLM) process to fabricate porous CoCrMo alloys. Their study aimed to determine the optimal volume porosities and heat treatment parameters to achieve a close match in elastic modulus and yield strength between human cortical bone and SLM-built CoCrMo alloys. Their results revealed a significant reduction in elastic modulus and yield strength with increasing actual porosity. Notably, heat-treated CoCrMo structures with an actual porosity of 48% demonstrated the most favorable mechanical properties, closely approximating those of human cortical bone, thereby presenting promising potential for biomedical implant applications.

Additionally, Jiang et al. [10] outlined an optimal process route for the preparation of Inconel 718 tools intended for cold, deep drawing applications. Utilizing Inconel 718 powder as the raw material, these authors optimized parameters for both Laser Powder Bed Fusion (L-PBF) and double annealing (DA), followed by surface finishing techniques. The resulting Inconel 718 tools exhibited a tensile strength of 1511.9 MPa and a hardness of 55 HRC. This process route successfully addressed the challenge of meeting the mechanical property requirements for cold deep drawing applications with L-PBF-built Inconel 718 tools.

2.2. Metal Matrix Composites

The thermal conductivity of metal matrix composites is a critical property for die production and electronic devices. Determining the effective thermal conductivity of additive-manufactured composites under various parameters often necessitates extensive experiments. Li et al. [11] addressed this challenge by constructing a theoretical model with a high accuracy of 86.7% for rapidly predicting the thermodynamic properties of laser-cladded Cu/Ni composites. The model's error was attributed to the ignorance of metallurgical bonding at the interface between the two different metals during the laser cladding process. It was observed that higher model accuracy could be achieved in samples with a larger cladding layer thickness. Meanwhile, Zhou et al. [12] employed their independently designed liquid–solid separation method to fabricate diamond/Al composites reinforced with 40 vol.% diamond particles, achieving high thermal conductivity. This innovative approach demonstrates the potential for producing composites with superior thermal properties for advanced industrial applications.

To ensure excellent interfacial bonding in Fe/Al composites post-heat treatment, a comprehensive understanding of the growth kinetics of intermetallic compounds (IMCs) is crucial. Zhang et al. [13] investigated the nucleation and early growth behavior of pure Fe/pure Al IMCs using in situ analysis. During the heat treatment process at 380 °C, the primary IMCs transitioned from the initial Fe_4Al_{13} phase to the stable Fe_2Al_5 phase. Initially, the growth rate of IMCs in thickness was closely aligned with the horizontal growth rate, ranging from 0.02 to 0.17 μm/min. However, upon reaching a thickness of 4.5 μm, the growth rate significantly decelerated to 0.007 μm/min. Similarly, during heat treatment at 520 °C, the predominant IMCs were of the Fe_2Al_5 phase, exhibiting a horizontal growth rate of 0.53 μm/min and a thickness growth rate of 0.23 μm/min. This investigation into the nucleation and early growth behavior of IMCs provides valuable insights, potentially reducing the reliance on costly trial-and-error processes in the microalloying of these composites.

The aerospace and transportation industries impose stringent requirements on the mechanical and damping properties of Al-based matrix composites (AMCs). Insufficient damping capacity has notably restricted their applications in vibration-sensitive environments. Lin et al. [14] investigated the mechanical and damping properties of SiC_f/Al-Mg composites with varying Mg contents. These composites were fabricated using colloidal dispersion combined with a squeeze-melt infiltration process. Their results indicated that the addition of Mg effectively enhanced interface bonding and strengthened the aluminum matrix, thereby improving the flexural strength and elastic modulus of the composite. However, higher Mg content led to the formation of pores, compromising the composite's plasticity. Furthermore, the incorporation of SiC fibers significantly enhanced the damping capacity of the composites, particularly under strain amplitudes exceeding 0.001%. This study provides valuable insights into achieving AMCs with superior mechanical performance.

As high-speed and heavy-haul rail transportation continue to evolve, there is an increasing demand for materials with a superior combination of hardness and toughness to prevent surface failures in rail turnouts. Zhao et al. [15] developed in situ WC primary reinforced bainite steel matrix composites with a notable hardness/toughness trade-off using direct laser deposition (DLD). These authors attributed the excellent balance between hardness and toughness to the adaptive adjustment of both the matrix and reinforcement microstructures, where a higher concentration of primary reinforcement contributed to enhanced mechanical properties.

3. Outlook

The present Special Issue attracted a substantial number of submissions, from which over 10 exceptional works were meticulously chosen to comprise the final publication. The Guest Editors express their sincere gratitude to all authors, reviewers, and publishers for their contributions and support of this endeavor. Encouraged by this success, a new Special Issue entitled "Additive Manufacturing of Alloys and Composites (Second Edition)" has been commissioned. Submissions from researchers worldwide are welcome for consideration in this forthcoming Special Issue, with further details available on the following website: https://www.mdpi.com/journal/materials/special_issues/NB0LNF40YT (accessed on 20 February 2024).

Funding: This work was sponsored by the Guangdong Basic and Applied Basic Research Foundation (No. 2022A1515012627) and Guangzhou Science and Technology Plan Project (No. 2024A04JB668).

Conflicts of Interest: The authors declare no conflicts of interest.

References

1. Herzog, D.; Seyda, V.; Wycisk, E.; Emmelmann, C. Additive Manufacturing of Metals. *Acta Mater.* **2016**, *117*, 371–392. [CrossRef]
2. Tan, C.; Weng, F.; Sui, S.; Chew, Y.; Bi, G. Progress and Perspectives in Laser Additive Manufacturing of Key Aeroengine Materials. *Int. J. Mach. Tools Manuf.* **2021**, *170*, 103804. [CrossRef]
3. Wang, J.; Wang, S.; Wu, W.; Wang, D.; Zhao, J.; Yang, Z.; Chen, H. Achieving Ultrahigh Yield Strength and Decent Toughness in an in-Situ Alloyed H13 Steel via Laser Powder Bed Fusion. *Addit. Manuf.* **2024**, *85*, 104169. [CrossRef]
4. Cheng, H.H.; Ma, H.W.; Pan, L.; Luo, X.; Liu, L.; Dong, H.K.; Song, T.; Wang, F.; Yang, C. Manufacturability and Mechanical Properties of Ti-35Nb-7Zr-5Ta Porous Titanium Alloys Produced by Laser Powder-Bed Fusion. *Addit. Manuf.* **2024**, *86*, 104190. [CrossRef]
5. Wang, Z.; Mao, P.; Huang, C.; Yu, P.; Li, W.; Yin, S. Deposition Mechanism of Ceramic Reinforced Metal Matrix Composites via Cold Spraying. *Addit. Manuf.* **2024**, *85*, 104167. [CrossRef]
6. Chang, G.; Zhang, X.; Ma, F.; Zhang, C.; Xu, L. Printing, Debinding and Sintering of 15-5PH Stainless Steel Components by Fused Deposition Modeling Additive Manufacturing. *Materials* **2023**, *16*, 6372. [CrossRef] [PubMed]
7. He, J.; Lv, J.; Song, Z.; Wang, C.; Feng, H.; Wu, X.; Zhu, Y.; Zheng, W. Maintaining Excellent Mechanical Properties via Additive Manufacturing of Low-N 25Cr-Type Duplex Stainless Steel. *Materials* **2023**, *16*, 7125. [CrossRef] [PubMed]
8. Gu, Y.; Lv, J.; He, J.; Song, Z.; Wang, C.; Feng, H.; Wu, X. Study On the Effect of Microstructure and Inclusions on Corrosion Resistance of Low-N 25Cr-Type Duplex Stainless Steel via Additive Manufacturing. *Materials* **2024**, *17*, 2068. [CrossRef]
9. Lam, T.; Chen, K.; Tsai, C.; Tsai, P.; Wu, M.; Hsu, C.; Jain, J.; Huang, E. Effect of Porosity and Heat Treatment on Mechanical Properties of Additive Manufactured CoCrMo Alloys. *Materials* **2023**, *16*, 751. [CrossRef] [PubMed]
10. Jiang, C.P.; Maidhah, A.A.; Wang, S.H.; Wang, Y.R.; Pasang, T.; Ramezani, M. Laser Powder Bed Fusion of Inconel 718 Tools for Cold Deep Drawing Applications: Optimization of Printing and Post-Processing Parameters. *Materials* **2023**, *16*, 4707. [CrossRef] [PubMed]
11. Li, Y.; Lin, C.; Murengami, B.G.; Tang, C.; Chen, X. Analyses and Research on a Model for Effective Thermal Conductivity of Laser-Clad Composite Materials. *Materials* **2023**, *16*, 7360. [CrossRef] [PubMed]
12. Zhou, H.; Jia, Q.; Sun, J.; Li, Y.; He, Y.; Bi, W.; Zheng, W. Improved Bending Strength and Thermal Conductivity of Diamond/Al Composites with Ti Coating Fabricated by Liquid–Solid Separation Method. *Materials* **2024**, *17*, 1485. [CrossRef] [PubMed]
13. Zhang, X.; Gao, K.; Wang, Z.; Hu, X.; Wang, J.; Nie, Z. In Situ SEM, TEM, EBSD Characterization of Nucleation and Early Growth of Pure Fe/Pure Al Intermetallic Compounds. *Materials* **2023**, *16*, 6022. [CrossRef] [PubMed]
14. Lin, G.; Sha, J.; Zu, Y.; Dai, J.; Su, C.; Lv, Z. Strengthening Mechanism and Damping Properties of SiC$_f$/Al-Mg Composites Prepared by Combining Colloidal Dispersion with a Squeeze Melt Infiltration Process. *Materials* **2024**, *17*, 1600. [CrossRef] [PubMed]
15. Zhao, M.; Jiang, X.; Guan, Y.; Yang, H.; Zhao, L.; Liu, D.; Jiao, H.; Yu, M.; Tang, Y.; Zhang, L. Enhanced Hardness-Toughness Balance Induced by Adaptive Adjustment of the Matrix Microstructure in In Situ Composites. *Materials* **2023**, *16*, 4437. [CrossRef] [PubMed]

MDPI AG
Grosspeteranlage 5
4052 Basel
Switzerland
Tel.: +41 61 683 77 34

Materials Editorial Office
E-mail: materials@mdpi.com
www.mdpi.com/journal/materials